D0122980

Is Einstein Still Right?

IS EINSTEIN STILL RIGHT?

CLIFFORD M. WILL
NICOLÁS YUNES

OXFORD
UNIVERSITY PRESS

OXFORD
UNIVERSITY PRESS

Great Clarendon Street, Oxford, OX2 6DP,
United Kingdom

Oxford University Press is a department of the University of Oxford.
It furthers the University's objective of excellence in research, scholarship,
and education by publishing worldwide. Oxford is a registered trade mark of
Oxford University Press in the UK and in certain other countries

© Clifford M. Will, Nicolás Yunes 2020

The moral rights of the authors have been asserted

First Edition published in 2020

Impression: 1

All rights reserved. No part of this publication may be reproduced, stored in
a retrieval system, or transmitted, in any form or by any means, without the
prior permission in writing of Oxford University Press, or as expressly permitted
by law, by licence or under terms agreed with the appropriate reprographics
rights organization. Enquiries concerning reproduction outside the scope of the
above should be sent to the Rights Department, Oxford University Press, at the
address above

You must not circulate this work in any other form
and you must impose this same condition on any acquirer

Published in the United States of America by Oxford University Press
198 Madison Avenue, New York, NY 10016, United States of America

British Library Cataloguing in Publication Data
Data available

Library of Congress Control Number: 2020937411

ISBN 978–0–19–884212–5

Printed and bound in the UK by
TJ International Ltd

Links to third party websites are provided by Oxford in good faith and
for information only. Oxford disclaims any responsibility for the materials
contained in any third party website referenced in this work.

For Leslie

For Penny, Jessica, Roberto and Lita

CONTENTS

PREFACE

A little over a century ago, over four consecutive Wednesdays in November 1915, Albert Einstein described to the Prussian Academy of Sciences a theory of gravity that he had been working on for eight years. Among the audience of German scientists, a few were excited and impressed, many were mystified, and some were openly hostile. Outside the world of German science, the lectures had almost no impact. This was the middle of World War I, and, except for a few neutral countries such as Switzerland and the Netherlands, Germany was effectively cut off from the rest of the world. Einstein, seriously ill from the grueling days and nights of calculating, and from the food rationing and other privations of wartime Berlin, returned to his office to continue toiling on his new theory in relative obscurity.

Just four years later, after British astronomers declared that Einstein was right about the Sun's gravity bending light, international headlines proclaimed Einstein to be the successor to Isaac Newton, the herald of a strange new universe governed by rubbery time, warped space, and mathematics so abstruse that only a handful of people could possibly comprehend it. Einstein became an overnight science superstar, a status that he thoroughly enjoyed and occasionally disliked. But his brainchild, called general relativity, soon languished, burdened by a shortage of relevance, a lack of experimental support and a reputation for being just too complicated. General relativity soon became little more than an afterthought in the world of physics.

But by 2015, the hundredth anniversary of general relativity, Einstein's theory had assumed its rightful place in the pantheon of physics. Its predictions had been tested and retested countless times, sometimes with mind-boggling precision. College bookshelves displayed textbooks on general relativity alongside conventional tomes on quantum mechanics, solid-state physics and astronomy, and physics departments routinely taught general relativity to graduate and undergraduate students. The theory's relevance was being touted in fields ranging from high-energy physics to astronomy to cosmology. And modern-day science superstars, such as Stephen Hawking, could be seen or heard expounding on warped spacetime on YouTube or in television shows such as *The Big Bang Theory*. It was even said that general relativity helps you to navigate your car or to find your misplaced smartphone, through the manner in which its rubbery time must be accounted for in global navigation systems such as GPS.

The crowning event of that centennial year was the September 14 2015 detection of gravitational waves emitted by a pair of colliding black holes a billion light years away from Earth. Einstein first predicted these waves in 1916, doubted their reality for a while in the 1930s, and believed that it would never be feasible to detect them. That detection, announced at a press conference in February 2016, made similar world-wide headlines proclaiming that Einstein was right. More importantly, it initiated a new way of doing astronomy, by "listening" to the universe rather than by looking at it. It also opened up new ways of putting Einstein's theory to the test, using black holes, neutron stars and gravitational waves.

This book is about Einstein's creation of over a hundred years ago, the general theory of relativity, with a definite slant toward experiment and observation. General relativity is a very beautiful theory. Einstein was guided toward its final form by aesthetic criteria of beauty, simplicity and elegance. In the end, while he appreciated the role of experimental tests, deep down he believed that the theory was so beautiful that it *had* to be correct. But, as the great American physicist Richard Feynman once said, "It doesn't matter how beautiful your theory is, it doesn't matter how smart you are. If it doesn't agree with experiment, it's wrong."

In this book we will describe how general relativity has passed every experimental test to which it has been subjected, an almost unbelievable perfect score. Yet the 1687 gravitation theory of Newton had a similar perfect score until general relativity took over. There is no reason to assume that general relativity is the last word on gravity. The observation of some anomalous effect or of a disagreement with Einstein's theory could tell us that it's time for a new theory. The 1998 discovery that the expansion of the universe is speeding up rather than slowing down is an example of an anomaly that has many people scratching their heads. Some of them are working hard on devising alternatives to general relativity to account for this. Therefore, we must keep testing general relativity, especially in new and unfamiliar arenas, such as near black holes, or using gravitational waves, in order to discover where or how, or even if, it might be superseded.

The authors are theoretical general relativists, but we have both spent a substantial fraction of our research careers investigating how to verify (or disprove) general relativity by experiment or observation. We don't actually do experiments or make observations; our experimental colleagues get nervous when we get too close to their equipment. Yet, we have spent enough time talking to them and collaborating with them that we think we have a good feeling for what they do and how observations and experiments can test Einstein's theory. In this book you will learn about some of the absolutely brilliant people who design the experiments, build the apparatus and instruments, and analyze the data. Some of them work alone or in small groups, some belong to enormous collaborations of thousands of scientists, engineers and technicians. These are the people who are doing the real work of finding out if Einstein is still right.

Acknowledgments

We are very grateful to the many friends and colleagues who read parts of this book and sent us criticisms, corrections and suggestions: Bruce Allen, Imre Bartos, Peter Bender, Donald Bruns, Alejandro Cárdenas-Avendaño, Katerina Chatziioannou, Ignazio Ciufolini, Karsten Danzmann, Sheperd Doeleman, Philip Eaton, Jim Hough, Cole Miller, Jenny Meyer, Paolo Freire, Reinhard Genzel, Ramesh Narayan, Jorge Pullin, Jessica Raley, David Reitze, Bernard Schutz and Norbert Wex. Ultimately, we are responsible for any remaining errors or omissions.

Clifford Will is grateful to the University of Florida for its support and to the Institut d'Astrophysique de Paris for its hospitality during extended stays in 2018 and 2019 while this book was being written. He is also grateful to the US National Science Foundation for support under various grants. Nicolás Yunes is grateful to Montana State University and the University of Illinois at Urbana-Champaign for their support, as well as to the Kavli Institute for Theoretical Physics for its hospitality during extended workshops in 2019 while this book was being written. He is also grateful to the US National Science Foundation and the US National Aeronautics and Space Administration for support under various grants.

In this book we will present "thought experiments" or situations where an observer does something. Rather than referring to the observer as "he or she" or "they", we will go back and forth, sometimes using "he" and sometimes using "she". This is by no means meant to exclude the possibility of observers or scientists who might be in gender transition or gender non-binary. Indeed, as physicists, we are well aware that our profession still needs to work hard to become more diverse in terms of gender, gender identification, race, ethnicity and disability status, and we are committed to doing our part. We hope that our usage of pronouns in this way will keep our examples readable and enjoyable, while still fostering a spirit of inclusiveness.

CHAPTER 1

A Very Good Summer

The summer of 2017 was a smashing success for Albert Einstein. On Thursday 25 May, in the waning weeks of spring that year, the journal *Physical Review Letters* posted a paper on its website. It described a nineteeen-year campaign to watch two stars revolving around the gargantuan black hole at the precise center of our own Milky Way. These two stars are special because their orbits are very close to the black hole, so they whirl around it at speeds as high as a few percent of the speed of light, or 20 to 30 million kilometers per hour. These orbits are ideal to test Einstein's theory and to search for any possible deviations from its predictions. No deviations were found by the team, headed by Andrea Ghez of the University of California in Los Angeles. Einstein passed yet another test, the first one to involve orbits around a black hole.

On Tuesday 18 July, at 8 p.m., standing in front of a forest of computer monitors in the German town of Darmstadt, a scientist at the mission operation center gave the kill signal. Five seconds later, 1.5 million kilometers from Earth, the LISA Pathfinder satellite shut down. A sigh of relief mixed with sadness was heard through the room. For sixteen months, two identical cubes of a gold and platinum alloy, 1.8 inches on a side, floated freely inside evacuated chambers in the satellite, maintaining almost exactly the same separation. The satellite had to periodically adjust slightly to account for shifts in its position caused by the bombardment of protons and radiation from the Sun. If either cube made contact with the walls of its chamber, it would be a disaster. Specially made spacecraft thrusters and delicate sensors were essential if this mission was to succeed. For sixteen months, the inside of this satellite was the quietest place in the universe. The success of the mission moved scientists one step closer to fulfilling a dream: the observation of gravitational waves with a space detector, known as LISA.

The month of August that summer was even better. On Monday 14 August the LIGO gravitational wave detectors in the USA and the Virgo detector in Italy picked up the signal from two black holes that merged 1.4 billion years ago. This wasn't the first gravitational wave signal detected—that momentous discovery had happened almost two years earlier—but it was the first to be detected simultaneously by the LIGO instruments in Washington state, near the Hanford nuclear reservation, and in Louisiana, near Baton Rouge, and by the newly operational Virgo detector near Pisa, Italy. The triple detection enabled the scientists to do a much better job of pinpointing the location of the source on the sky.

Three days later, another burst of gravitational waves jiggled the sensitive mirrors of the LIGO and Virgo detectors. A few seconds after that, the Fermi Gamma-Ray Space Telescope, orbiting 534 kilometers above the Earth, sensed a burst of gamma rays coming from the same part of the sky. Some rapid detective work located the galaxy where both signals originated, and during the subsequent hours and days astronomers around the world observed light in all its forms, from X-rays to radio waves, arriving at Earth from that same location. This time, the source was two neutron stars about 140 million light years from Earth producing waves of gravity as the stars spiraled toward each other and merged. This was followed by a nuclear fireball of unimaginable power.

This single observation revealed wonders about the universe that not even Einstein would have imagined. If you are wearing a gold necklace or a platinum ring, then there is a good chance that those precious (and expensive) elements were produced in a nuclear cataclysm just like the one observed on that day. In fact, most of the gold and platinum in the universe is now thought to have been produced in the explosions that result when neutron stars collide.

If that wasn't enough, the mere fact that the gravitational waves, emitted right before the neutron stars merged, and the first gamma rays, emitted right after the merger, arrived within 2 seconds of each other after traveling 140 million light years revealed that the speed of gravitational waves and the speed of light are the same to fifteen decimal places. Amazingly, this is precisely what Einstein had predicted in 1916.

The following week, on Monday 21 August, a lone amateur astronomer named Don Bruns settled into a folding patio chair near the top of Casper Mountain in Wyoming. With the press of a key, his laptop began instructing a TeleVue Optics NP101is telescope to take a series of photographs of the Sun during the total solar eclipse that crossed the United States that day. His goal was to replicate a famous 1919 experiment carried out by a British team of professional astronomers headed by Arthur Stanley Eddington. Eddington's measurements showed that gravity bends light exactly as Einstein had predicted, thereby overturning Newton's theory and making Einstein an international celebrity. Bruns wanted to see what could be done by a non-professional astronomer armed only with a modern commercial telescope, a CCD (charge-coupled device) camera, and computer-controlled instruments. After analyzing his data Bruns also verified that light is bent by the Sun just as Einstein predicted, and the accuracy of his measurements beat Eddington's by a factor of three.

Many of these events were covered by the press, using headlines like "Einstein was right, again." They reinforced an almost fairy-tale version of the story of general relativity that goes something like this: in 1905, Einstein, working as a lowly clerk at the Patent Office in Bern, Switzerland, created the Special Theory of Relativity. He then turned his attention to gravity, and after ten years of hard work created the General Theory of Relativity. In 1919 Eddington verified the theory by measuring the bending of starlight. Einstein became famous, his theory was triumphant, and everybody lived happily ever after.

The actual story of general relativity is more complicated. Back in the 1920s there was considerable skepticism about Eddington's results, particularly among American astronomers. Attempts in 1917 to measure the shift in the wavelength of sunlight toward the red end of the spectrum, an effect that Einstein considered another crucial test of his theory, *failed* to detect the effect. This apparently hurt Einstein's chances for the Nobel Prize until 1921, when it was finally awarded for his work on the photoelectric effect, not for general relativity.

The theory was considered to be extremely complicated, with exotic new concepts like curved spacetime that baffled most physicists and astronomers at the time, to say nothing of the general public. The headline of an article on relativity in the 9 November 1919 issue of the *New York Times* stated, "A book for 12 wise men / No more in all the world could comprehend it, said Einstein when his daring publishers accepted it." Einstein himself may have used some such phrase as early as 1916 in reference to a popular book on relativity that he had written. Another story has this idea originating with Eddington. Soon after the publication of the final form of general relativity in 1916, Eddington was one of the first to appreciate its importance, and set out to master the theory and then to organize a team to measure the bending of light. At the close of the November 1919 joint meeting of the Royal Astronomical Society and the Royal Society of London at which Eddington reported the successful measurements, a colleague purportedly said, "Professor Eddington, you must be one of three people in the world who understand general relativity!" to which Eddington demurred. The colleague persisted, saying, "Don't be modest, Eddington." Eddington replied, "On the contrary, I am trying to think who the third person could be."

Perhaps only a handful of people understood it, but millions were fascinated by it and wanted to read about it and about Einstein. In the popular press, the scientific revolution engendered by general relativity was placed on a par with the insights of Copernicus, Kepler and Newton. Editorial after editorial marveled at what was called one of the greatest achievements in the history of human thought, but at the same time complained about the difficulty of understanding it. Einstein himself wrote a long article for *The Times* of London in late 1919, attempting to explain the theory to a general audience. His picture graced the cover of the 14 December 1919 issue of the German news magazine *Berliner Illustrirte Zeitung*, with the caption "A new great figure in world history."

But there was a sense among scientists that the only thing this complex theorizing was good for was to predict minute deviations from Newton's grand theory. Experimentalists held sway in the physics of that time, and they felt that general relativity would never play a role in mainstream science.

As a result of this skepticism, the science of general relativity gradually became stagnant and sterile. By the mid 1920s Einstein had turned most of his attention to what would become a futile quest for a unified field theory that would combine gravitation and electromagnetism, and many other relativity researchers followed suit. With only a few exceptions, most work in general relativity during the next thirty-five years was devoted to abstract mathematical questions and issues of principle, and was carried out by a small band of practitioners. Science historian Jean Eisenstaedt has called this period the "low water mark" for Einstein's theory. A classic illustration of the attitude toward Einstein's theory at the end of this period is the advice given in 1962 to a newly minted graduate of the California Institute of Technology, about to head for Princeton to do graduate work. A famous Caltech astronomer advised him that he should absolutely *not* work on general relativity when he got to Princeton, because it would *never* have anything useful to contribute to physics or astronomy. Luckily for many of us, the student, Kip Thorne, ignored the advice.

As Kip headed east for the ivy covered walls of Princeton, Caltech astronomers were on the verge of announcing the discovery of strange objects that they called "quasistellar radio sources," or quasars. These were very distant, *very* energetic sources of radio waves that defied explanation in terms of conventional physics. A few people started to wonder if general relativity might help to provide an explanation, and convened a special conference on quasars, bringing together astrophysicists and general relativists. Held in Dallas in December 1963, this historic conference would come to be known as the First Texas Symposium on Relativistic Astrophysics. Within a few years, other discoveries pointed to a definite role for general relativity in astrophysics. In 1965 came the detection of the cosmic background radiation left over from the big bang. In 1967 radio astronomers discovered the first of many pulsars, now understood to be rapidly spinning neutron stars. And 1971 saw the discovery of a compact, powerful X-ray source orbiting a normal star, the first black hole candidate. You needed general relativity if you wanted to begin to understand these phenomena.

These discoveries helped bring about a renaissance for general relativity, in which it would begin to rejoin the mainstream of physics and astronomy. This was aided by advances in technology, such as atomic clocks, lasers and superconductors, and by the development of the space program, which would provide the tools to perform new high-precision tests of Einstein's theory, putting it on a solid experimental foundation. After all, if you needed to use a relativistic theory of gravity to understand quasars, pulsars and the cosmic background radiation, it would be very good to know if Einstein's theory was the correct one to use. For despite Einstein's fame, by this time there were alternative theories of gravity competing with general relativity for primacy. One of these, known as the Brans–Dicke theory, named after Princeton University's Robert Dicke and his student Carl Brans, made a credible claim that it was just as viable as Einstein's theory. This competition sparked a major effort to carry out new and better experimental tests to determine if Einstein was right or not.

Other factors helped set the stage for this rebirth of general relativity research. Ironically, one of these may have been the death of Einstein himself in April 1955. No other topic in physics was so closely tied to a single, towering individual. It was not uncommon in those days for the few practitioners of general relativity to make a pilgrimage to Princeton to describe their work to the great one, and, hopefully, to receive his approval. On the occasion of the centenary of general relativity in 2015, the French mathematical physicist Yvonne Choquet-Bruhat wrote a charming reminiscence of her visit to the Institute for Advanced Study in 1951 as a 27-year-old postdoctoral researcher. She visited Einstein several times during that year, explaining her mathematical work on the existence of solutions of Einstein's equations, and listening to Einstein describe his work on unified field theory. He pronounced himself very pleased with her work, which actually turned out to be one of the major milestones of the field. His work on the unified theory ultimately went nowhere. But after Einstein passed from the scene, the field was somehow free to make its own way forward.

Another factor may have been an emerging sense of community among the small group of specialists in relativity. Beginning with a conference in July 1955 in Bern, Switzerland to commemorate the fiftieth anniversary of special relativity, regular international meetings on the subject were organized. By 1959, at the third such meeting, held in Royaumont, France, leaders of the field formed the "International Committee" on general relativity, to aid the organization of future meetings, to disseminate lists of published papers in the field and to provide information on research groups around the world. This would ultimately evolve into the International Society on General Relativity and Gravitation, with elected officers, annual dues and its own scientific journal.

The fact is that science is more than a body of knowledge. It is also a community of researchers who advance the field by sharing knowledge, collaborating and competing with one another, and even correcting each other, so that scientific facts can be established and progress can be made. Some historians of science believe that the emergence of this community of relativists in the late 1950s made it possible for them to respond nimbly and effectively to the new astronomical discoveries of the 1960s.

By the time of the 1979 centenary of the birth of Einstein, the relativistic renaissance that had begun in the 1960s was in full swing. The outpouring of books commemorating that birthday attested to the vigor and excitement of research in the field. A "toolbox" of techniques for solving Einstein's complicated equations in a wide range of situations had been developed, and researchers were turning to computers to help solve the more complex problems. Many experimental tests of general relativity had been performed. Some exploited new technologies to do improved versions of "classic" experiments, such as measurements of the bending of light using radio waves from quasars rather than using visible light from stars. Others were new tests never envisioned by Einstein, such as the "Shapiro time delay," an excess delay in the propagation of radar signals as they passed near the Sun on their way to track the orbits of planets or spacecraft.

Black holes were an accepted and understood part of the theory, and observational evidence that they actually exist was mounting. A model for the basic structure and evolution of the universe was in good shape, and cosmologists were beginning to explore what might have happened in the first trillionth of a second after the big bang. The nature of quasars ironically remained a mystery after twenty years of study, while a pretty good theory of pulsars was in hand.

The centenary year itself was inaugurated by a stunning announcement. At the ninth Texas Symposium on Relativistic Astrophysics, held in Munich in December 1978, Joseph H. Taylor of the University of Massachusetts described the latest result from a remarkable new testing ground for general relativity that he and his student Russell Hulse had discovered in 1974. This new arena for experimental relativity was a pulsar in orbit about a companion star, colloquially called the "binary pulsar." Taylor reported how observations of the orbit of the pulsar since 1974 had led to the first confirmation of one of the most important predictions of Einstein's theory: the existence of gravitational waves. Nineteen years later, Taylor and Hulse would win the Nobel Prize in Physics, the first ever bestowed for work related to general relativity. If general relativity *seemed* to be triumphant in 1919, it was surely triumphant now.

But there were some clouds on the horizon, and there would be more to come. None of them posed a direct threat to the supremacy of Einstein's theory—as of 1979 it had not failed a single experimental test, and that perfect record continues to today—but they suggested the possibility that Einstein might not have had the final word on gravity.

The first cloud was theoretical. In addition to developing special and general relativity, Einstein was one of the pioneers of quantum mechanics. And although Einstein ultimately found himself at odds with some of the interpretations of quantum physics, he would have been the first to admit that it was spectacularly successful in its ability to account for measurable effects in the subatomic world. And even some of the "spooky" probabilistic effects that he frequently railed against ("God does not play dice!") have been shown by recent experiments to be true and unavoidable. Quantum mechanics rules everything from the practical (chemistry, semiconductors, MRI, nuclear energy, the workings of your smartphone, the list is almost endless) to the exotic (quarks, the Higgs boson, quantum computers, ...). The basic forces of the inner world—the strong force, which governs atomic nuclei, the electromagnetic force, which governs charged particles and light, and the weak force, which governs some forms of radioactive decay and is intimately linked to the elusive particles called neutrinos—are all understood today using quantum mechanics. This notion that "the quantum" rules everything is so pervasive in physics that it is almost an act of faith that the weakest of the forces, gravity, must also somehow be "quantized."

It is more than just faith, however. When treated as a pure, exact theory, general relativity actually seems to sow the seeds of its own destruction. The most common example of this is the black hole, which, as we will see later in this book, contains a "singularity" at its center. A singularity is a point in space where the warpage of spacetime, the density, the pressure and the energy all become infinite. Similarly singular behavior occurred at the instant of the universal "big bang," according to standard general relativity. But the appearance of infinities in the predictions of a theory has frequently been interpreted as a signal that the theory itself is breaking down. For example, when experiments carried out in 1911 indicated that the atom consists of a tiny positively charged nucleus surrounded by orbiting negatively charged electrons, scientists realized immediately that there was a problem. The revolving electrons should emit electromagnetic radiation, lose energy and spiral inward toward the nucleus, emitting along the way a potentially infinite amount of energy. This was theoretically untenable, and of course it also contradicted experiment. The quantum mechanical model of the atom devised by Niels Bohr came to the rescue by requiring that electrons stay on fixed orbits. They would only radiate light when they made a "transition" or a jump from one orbit to an adjacent orbit of lower energy. And every atom had a "ground state," an orbit of lowest energy, from which no further jumps were permitted. The more fully developed quantum mechanics of Erwin Schrödinger and Werner Heisenberg refined this picture in terms of probabilities and the uncertainty principle.

In the case of the black hole or big bang singularities, the hope was that quantum mechanics might come to the rescue in a similar way, for example by somehow preventing things from falling all the way to the black hole singularity, the way the atomic ground state of an atom keeps the electron from falling all the way into the nucleus.

But therein lies the rub. After almost a hundred years we still do not have an acceptable quantum theory of gravity. This is not for a lack of trying, and there is a sizable, worldwide research effort attacking this problem from a dizzying array of directions, with arcane names like canonical quantum gravity, superstring theory, causal set theory, loop quantum gravity and the AdS/CFT correspondence, to name just a few. We won't burden you with a list of reasons why this problem is so difficult, but instead we will skip to the bottom line: general relativity as Einstein formulated it in 1915 cannot be the quantum theory of gravity. Whatever that theory turns out to be, it will be different from general relativity. The questions are how different, and can we ever detect those differences? One viewpoint asserts that quantum gravity becomes relevant only between the big bang and about a tenth of an atto-yottosecond (or 10^{-43} seconds) later, or at energies a million billion times higher than the levels achieved by the Large Hadron Collider in Geneva. Inside a black hole, quantum effects would kick in only at similarly incredibly short distances from the singularity, and in any case, since the event horizon that surrounds the singularity prevents the escape of any information about what is going on, why would we care? Consequently, according to this viewpoint, we will *never* be able to perform an experiment capable of detecting a quantum gravity effect. This raises a scientific conundrum. If you have two theories and there is no conceivable, practical way ever to perform an experiment to test them, how do you decide which one is correct? Elegance? Simplicity? A democratic vote? Faith? And is that science?

Our viewpoint on quantum gravity is agnostic. As you will discover in this book, we will be mostly concerned with experiments that can be performed in a finite amount of time, with a finite (if sometimes rather large) budget. As a result, the book will not contain a detailed description of the different attempts to develop a theory of quantum gravity, a topic with a long and complicated history of its own. We must admit, however, that the scientific community knows so little about how to make general relativity compatible with the quantum that it is entirely possible that some future experiment might stumble on an effect in conflict with the predictions of general relativity. If so, such an experiment or observation could point the way forward toward a theory of quantum gravity. For now, we can only continue to test Einstein's theory to higher precision and in new regimes.

Another cloud on Einstein's horizon is called dark matter. Most physicists and astronomers are now convinced that only about 4 percent of the mass of the observable universe is made up of the normal matter that we know and love because we are made of it: protons, neutrons, electrons and other elementary particles that comprise what physicists call the Standard Model. About 23 percent is called "dark matter." (We'll get to the remaining 73 percent in a moment.) The evidence supporting dark matter is compelling. The velocities of stars and gas around spiral galaxies, the velocities of galaxies within clusters of galaxies, and the bending of starlight around galaxies and clusters are all too large to be explained by the mass of ordinary matter making up the visible galaxies themselves. There is additional mass around these objects that does not emit light, but that attracts gravitationally, hence the name "dark" matter. Observations of the pattern of tiny fluctuations in the intensity of the cosmic background radiation confirm the presence of dark matter. Even the formation of galaxies themselves, beginning about a million years after the big bang, cannot be properly understood unless there is dark matter, using its gravitational attraction to tug the ordinary matter into the blobs that ultimately collapsed and coalesced to form the stars and galaxies that we observe.

A leading candidate for dark matter is an elementary particle that is not part of the Standard Model; by suitably tweaking that model, particle theorists have come up with numerous plausible candidates. If so, millions of these particles should be passing through your body every second as you read this book, and thus a sensitive enough detector ought to be able to sense some of them. But despite almost forty years of experimental effort carried out by physicists around the world, no detection has been made. To some, this has become a bit of an embarrassment, and so they have turned to an alternative approach: why not modify gravity itself? Although, as you will discover in this book, general relativity has been confirmed in many different arenas, it has not been well tested on the very large distance scales of galaxies, clusters of galaxies, or the observable universe as a whole. So perhaps a suitable modification of general relativity would solve the dark matter problem. So far, while most of these "modified gravity" theories have not been very successful, a future theory could be, with its predictions confirmed by future observations, especially ones that focus on these large distance scales.

What is the remaining 73 percent? Enter the final cloud on the horizon: "dark energy." In 1998, astronomers studying very distant supernova explosions were forced to conclude, from the pattern of their data, that the expansion of the universe is speeding up, not slowing down. Given that we have known since 1929 that the universe is expanding, your intuition is perhaps telling you that the expansion should be slowing down. After all, mass makes gravity and gravity attracts, so, just as the Earth's gravity causes a ball thrown in the air to slow down and come back down, the universe should make its own expansion slow down (whether it continues to expand but ever more slowly or halts and starts to contract is a separate question). And indeed, Einstein's general relativity predicts unambiguously that the expansion of the universe should slow down or decelerate, not speed up. So when this *accelerated* expansion was detected, it was a major shock.

But then theorists got to work coming up with ideas to account for this new phenomenon. One set of ideas was dubbed "dark energy" by University of Chicago cosmologist Michael Turner, in analogy with dark matter—the adjective is rather prophetic, because we are still pretty much in the dark about both of them. If you ascribed to this substance the right gravitationally repulsive or "anti-gravity" properties, and if you let it contribute about 73 percent of the total mass and energy in the observable universe, then it fits all the observational data beautifully. In fact the cosmological model based on this, called the "Lambda-CDM" model, has proven to be remarkably successful in explaining a diverse array of data on the universe at the largest scales. Here, the Greek capital letter lambda (Λ) refers to dark energy, while CDM refers to cold dark matter. But trying to understand what dark energy is from deeper principles of quantum mechanics and particle physics has led to a plethora of competing models, which are very hard to distinguish among through experiment or observation.

Another idea was to resurrect what Einstein called his "greatest blunder." Already in 1916 he was thinking about applying his new theory to the universe as a whole. But to his horror, he realized that the equations demanded that the universe must either expand or contract. It could not be static. At the time, conventional wisdom held that the universe was actually static, perfectly unchanging. In fact, it was not even known that there were galaxies outside the Milky Way. To get around this, he added what he called a "cosmological term" to his original equations. This term would have the effect of introducing a repulsion that would counteract the natural tendency of the universe to contract under its own gravitational attraction, leading to a nicely balanced, static universe. The size of the term was governed by a "cosmological constant," which came to be denoted by the Greek letter Λ (in homage to Einstein, dark energy aficionados have adopted the same symbol). So by picking the "right" value for his cosmological constant, Einstein could make everything right with the consensus view of the universe.

But along came data to mess with Einstein's world view. First was the discovery of galaxies external to our own, along with evidence that many of them were moving away from us. And then, Edwin Hubble, an astronomer at the Mount Wilson Observatory in California, announced in 1929 that the data on the motions of galaxies implied that the universe was not static but was expanding. Einstein was now forced to drop his cosmological term, since its sole purpose was to make the universe static. With the new knowledge that the universal expansion is accelerating, it is a simple matter to bring back Einstein's cosmological term, since it naturally provides the repulsive effect needed to counteract the gravitational deceleration. The value of the cosmological constant needed to explain the acceleration of the universe is much smaller than what Einstein needed for a static universe, and so the effect of adding this term to his equations is utterly negligible for everything but cosmology itself. You might call this the "minimal" modification to Einstein's theory.

A third idea is to modify general relativity more drastically, but still with enough fine tuning so as not to violate the agreement with the many experiments that we will be describing in this book. This turns out to be not so easy. In developing general relativity, Einstein was driven by a desire for elegance and simplicity in the structure of the theory, and he was remarkably successful. For all the talk about how complicated his theory appeared at the time of Eddington's announcement, from a more modern perspective, general relativity is the simplest theory of gravity you could possibly imagine. And it turns out that efforts to modify it for cosmological purposes generally lead to very ugly, complicated theories. Of course, it is not clear that the true theory of nature needs to be elegant or beautiful, since after all, these are human concepts that may have nothing to do with the physical world. In other words, the universe is a messy and dirty place, so it may be that the correct theory to describe it is equally messy and dirty.

None of these "clouds," quantum gravity, dark matter or dark energy (and others we don't have space to discuss), directly invalidates general relativity. Still, they leave us with the disquieting feeling that it might be necessary to take gravity "beyond Einstein"; to develop a theory that agrees with general relativity in all the realms where it has been tested precisely, but that might deviate from it, either in the realm of ultra-short scales and ultra-high energies, as in quantum gravity, or in the realm of ultra-large scales as in cosmology.

This book will focus on the many precision experiments and observations that have been carried out to test Einstein's theory in different realms (laboratory, solar system and astrophysical). But after reading about how general relativity has passed test after test, you might be tempted to say "Einstein is still right," so let's be done. However, in science in general, and in physics specifically, the acceptance of any theory is always provisional, because no theory can be fully tested in all possible realms of its applicability to perfect or infinite precision. The best we can do is to extend our experimentation into wider and wider realms and to higher and higher precision, in the hope of either building further confidence in the theory, or finding a deviation that might lead us to a new, more fundamental, and more complete theory. The history of science is full of examples of both outcomes.

For general relativity, the arena for experimental tests began with the solar system, notably with the famous measurement of light bending in 1919. By the 1970s, the arena had been extended to astrophysical scales, with the discovery of the binary pulsar. But since the 1970s, the precision to which tests can be performed in both of these realms has greatly improved, and completely new arenas have opened up, to include gravitational waves and black holes, for example. The events of Einstein's wonderful summer of 2017 illustrate all these arenas for putting general relativity to the test.

CHAPTER 2

Wrinkles in Time

Midway through the movie *Interstellar*, the crew of the *Endurance* discuss how to explore Miller's planet, which orbits just outside the event horizon of the supermassive black hole Gargantua:

COOPER (Matthew McConaughey) Look, I can swing around that neutron star to decelerate …

BRAND (Anne Hathaway) It's not that, it's time. That gravity will slow our clock compared to Earth's. Drastically.

COOPER How bad?

ROMILLY (David Gyasi) Every hour we spend on that planet will be maybe … seven years back on Earth.

COOPER Jesus …

ROMILLY That's relativity, folks.

Until the twentieth century, it was accepted by everybody that time was the universal time of Newton, flowing at the same rate everywhere and forever, the same for everybody. Einstein upended that comforting absolutism with his 1905 special theory of relativity, pointing out that time could flow at different rates for people who are moving relative to each other. And this is not just a funny effect of clocks. It is actual physical time that flows at different rates, making people age differently. And in 1911 he argued that gravity could also affect time in a similar way.

This striking prediction was based on what Einstein later called his "happiest thought." It was 1907, eight years before the general theory of relativity, and Einstein was doing well. The five papers he had published in 1905 on the photoelectric effect, the quantum nature of light, Brownian motion, special relativity and the mass–energy equivalence were generating a lot of buzz, and he would soon leave the patent office in Bern, Switzerland, where he had been working for the last six years, for a faculty position at the nearby University of Bern. He had also been invited to write a review article on special relativity for the scientific journal *Jahrbuch der Radioaktivität und Elektronik*. Part of the article would be a standard review of his special relativity theory, plus work that he and others had done on it since 1905, but part of it would be devoted to his latest preoccupation: gravity.

For all its successes, special relativity had a weakness. It was based on the premise of the "inertial reference frame," a laboratory that moves with constant speed in the same direction. Inside this laboratory, free particles (those not acted upon by forces such as electric or magnetic forces) move on straight lines at a constant speed. By analyzing the laws of physics within such frames of reference, Einstein was able to account in a natural way for the experimental fact that the speed of light was independent of the speed of the emitter or of the observer. He could also make sense of the interrelations between electric and magnetic fields as seen by observers in relative motion. His theory of special relativity had led to some surprising, and at that time untested, predictions, such as the idea that a moving clock would tick more slowly than clocks at rest, and that energy and mass were really the same thing, related by his famous $E = mc^2$ equation, destined for T-shirts and coffee mugs everywhere.

It was not too difficult to imagine an inertial reference frame, such as a spaceship with its engines turned off, in outer space far from any stars or galaxies, but what about on or near the Earth? Any freely moving particle would experience an acceleration, a change of its speed and of the direction of its motion, because of Earth's gravity. No reference frame could be truly inertial in the presence of gravity. Already in 1907, Einstein realized that he would have to find a way to make special relativity compatible with gravity.

It was here that Einstein demonstrated his special genius for taking a simple experimental observation, combining it with an idealized imaginary experiment (called a *gedanken* or "thought" experiment) that incorporates the essence of the original experiment, and pushing the result to its logical limit. In this case, the observational result was the commonplace one that bodies fall with the same acceleration, regardless of their internal makeup, in the absence of air resistance. This brings to mind the image of Galileo Galilei dropping objects from the top of the Leaning Tower of Pisa, although there is no actual contemporary account of his ever doing such a thing. But by the turn of the twentieth century this observational result had been verified to a few parts in a billion by a Hungarian physicist, Lorand Eötvös. Einstein took this simple observation and imagined what it would imply for an observer inside an enclosed, freely falling laboratory.

Of course, in 1907, when Einstein first began to ponder this question, it had to be a pure thought experiment, for the dawn of the space age and of astronauts floating around their space capsules was still fifty years into the future. There is also a story that he once observed a worker falling from a roof, and began to imagine what it would be like to be weightless (forgetting about what would happen when the worker encountered the ground, which from the worker's point of view would be rising ever more rapidly toward him!). Nevertheless, the weightlessness, or vanishing of gravity, that such an observer would experience seemed so significant to Einstein that he elevated it to the status of a principle. He called it the principle of equivalence.

"Equivalence" came from the idea that life in a freely falling laboratory should be equivalent to life with no gravity. It also came from the converse idea that, if you were in an enclosed rocket with no windows, far away from any star or galaxy, and accelerating with the right amount (called "one g") of constant rocket thrust, then you would not be able to tell that you are inside a rocket, as opposed to being safely inside a building on Earth. To Einstein, the acceleration due to a rocket and an acceleration due to gravity were the same thing! From this principle of equivalence, Einstein was able to conclude that time at the top of a tower at rest in a gravitational field ticks a little more quickly than time at the bottom.

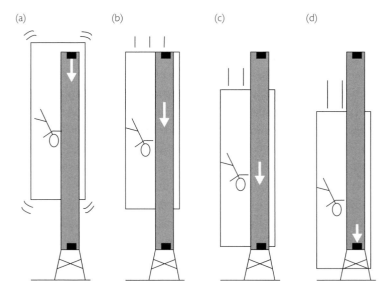

Fig 2.1 Gravitational redshift thought experiment. A laboratory is released at the moment the emitter sends the pulse of light (left panel). The observer inside can use the gravity-free laws of special relativity to analyze the emission, propagation and reception of the pulse. Second panel: the laboratory has started to fall downward, but because the observer inside senses no gravity, she sees the packet of light propagating with the same frequency as before. Third panel: the laboratory is falling faster, and the observer sees the receiver coming up toward her. Because the light packet still has the same frequency as seen by her, the ascending receiver will see a higher frequency (a blueshift), because of the Doppler shift (fourth panel). The velocity that determines the amount of the shift is just the speed that the laboratory picked up in the time it took for the light packet to go from the emitter to the receiver.

The easiest way to understand this is to consider a simple thought experiment involving the Earth, an emitter and receiver of light, and a freely falling laboratory (Figure 2.1). This thought experiment is a somewhat modernized version of what Einstein wrote in a 1911 paper, but the idea is the same. Imagine a device that emits light at a well-defined frequency or wavelength, placed at the top of a high tower with its beam directed downward. A tunable receiver is placed on the ground and is tuned to receive the incoming signal from the emitter on the tower. Relative to the emitted frequency, is the received frequency larger, smaller, or the same?

To answer this question, let's imagine a laboratory that is suspended next to our tower by a mechanism that can release the laboratory in an instant, letting it fall freely toward the ground. For a more visceral idea, picture being at the top of the 200-foot vertical drop of a roller coaster ride, such as "SheiKra" at Busch Gardens in Tampa, Florida, where the cars stop momentarily before being released, followed by a few seconds of breathtaking vertical free fall (and screaming). In this ride, the roller coaster obviously follows the curved tracks before hitting the ground, while in our thought experiment, the lab does crash against the ground, although this is irrelevant to our physics problem.

Let us also imagine that the laboratory is released at the *precise* moment that a short pulse or "packet" of light is also released by the emitter. Inside the laboratory, an observer prepares to measure the frequency of the emitted packet of light as she falls freely toward the ground. Clearly, the wave packet will travel toward the receiver at the speed of light, therefore faster than the laboratory and the observer are falling. The packet will thus overshoot the observer and reach the receiver before the laboratory hits the ground, as depicted in the four panels of Figure 2.1.

Now, let us think about what the observer measures at each stage of her descent. At the very start of the drop (the first panel in Figure 2.1), the laboratory is initially at rest with respect to the emitter, if only for a moment, and so the emitted frequency is the "rest" frequency, unaffected by any slowing down of moving clocks predicted by special relativity. The measured frequency would therefore be the standard value for that emitter, and could be looked up, say, in standard tables of physical constants, or calculated using the standard laws of atomic or nuclear physics.

At the next instant (the second panel of Figure 2.1), the wave packet travels down while the laboratory and observer begin to fall because of the Earth's gravitational pull, so what does the observer measure now? She realizes that throughout her descent she is in free fall, and because she is well versed in the principle of equivalence, she also realizes that, from her point of view, gravity is absent! The wave packet thus obeys the laws of special relativity, which state that light moves at a constant speed with an unchanging frequency. She thus measures that the frequency of the wave packet remains unchanged, as seen by her, during every stage of her fall.

A little while later, however, the observer notices that from her viewpoint the ground, and in particular the receiver, are coming up toward her! She is falling, so of course this is what she will experience, even though from the viewpoint of a person standing on the ground the receiver is clearly not moving (the third panel of Figure 2.1). Thus, from our observer's viewpoint, when the onrushing receiver absorbs the packet of light (the fourth panel of Figure 2.1), it will measure a higher frequency than our observer measured in the freely falling laboratory because of the Doppler effect, the effect that causes the frequency or pitch of an ambulance siren to be higher when the ambulance is approaching you and lower when it is moving away from you. And that is the answer to our question! Relative to the emitted frequency, the received frequency is higher.

The emitter and receiver, of course, are still at rest with respect to each other, but this is not the point. The important point is that *from the point of view of the observer* in the freely falling laboratory, in which the frequency has its standard value, the receiver is moving toward her. The velocity of the laboratory relative to the receiver is the same as the velocity that the freely falling laboratory has picked up in the time taken for the wave packet to travel the distance between the emitter and the receiver, and from this one can calculate the shift in frequency of the light. For example, for a difference in height of 100 meters, the shift would be only ten parts in a million billion, or one trillionth of a percent! If the emitter and receiver are at the same height, but separated in the horizontal direction, there is no frequency shift at all.

In this thought experiment the observed shift was toward higher frequencies—the blue end of the visible spectrum—because the freely falling frame was heading toward the receiver. If the emitter had been at the bottom and the receiver at the top, the shift would have been toward lower frequencies—the red end—because by the time the wave packet reached the top, the freely falling frame would be falling away from the receiver. Even though the result can be either a redshift or a blueshift, depending on the experiment, the generic name for this effect is the gravitational redshift. It is called a "gravitational" shift because it occurs only in the presence of a mass (the Earth in our case) that exerts a gravitational force on the lab (forcing it to accelerate down in our example).

It should be apparent from our thought experiment that the gravitational redshift is a truly universal phenomenon. It was the behavior of the freely falling laboratory that was the crucial element in the analysis. The nature of the emitter and receiver did not play a significant role, nor did our treatment of the nature of light. The light could have been in the visible spectrum, or it could have been in the radio or X-ray wavelengths. The signal could have been a continuous beam, or it could have been in the form of packets, such as might be emitted by a strobe light set to flash once per second. In the latter example, the observer at the bottom of the tower would observe not only that the intrinsic frequency of the light emitted by the strobe was shifted toward the blue, but also that the flashes arrived more quickly than once per second. Thus, all frequencies appear to be shifted. If the strobe's flashes were timed by some sort of clock, then the observer on the ground would argue that the clock at the top of the tower was ticking faster than his ground clock; in other words, that the clock rate was "blueshifted."

In fact, the distinction between clock and emitter/receiver of light that we have used is purely a semantic one. The term "clock" really means a device that performs some physical activity repetitively at a well-defined, constant rate. The activity could be the mechanical sweep of a second hand, the flashes of a strobe or the waves of an electromagnetic signal. Modern atomic clocks are based on the latter phenomenon—the emission of light with a constant, stable, well-defined frequency. The gravitational redshift affects all clock rates equally; this includes biological clocks, since, after all, biological processes fundamentally involve atoms and molecules, which are governed by the laws of physics. This can all be summed up in the simple statement that gravity warps time.

Another thing that should be apparent is that we did not use general relativity itself anywhere in the discussion. The gravitational redshift depends only on the principle of equivalence. Even though the full version of the general theory predicts the redshift effect, and Einstein viewed the redshift as one of the three main tests of his theory, we now regard it as a test of the more fundamental equivalence principle. Any theory of gravity that is compatible with the equivalence principle (and there are many, including, for instance, the theory by Brans and Dicke) automatically predicts the same gravitational redshift as general relativity.

A question that is often asked is: Do the intrinsic rates of the emitter and receiver or of the clocks change, or is it the light signal that changes frequency during its flight? The answer is simple: it doesn't matter! Both descriptions are physically equivalent. Put differently, there is no way to carry out an experiment to distinguish between the two descriptions. Suppose that we tried to check whether the emitter and the receiver agreed in their rates by bringing the emitter down from the tower and setting it beside the receiver. We would find that indeed they agree. Similarly, if we were to transport the receiver to the top of the tower and set it beside the emitter, we would find that they also agree. But to get a gravitational redshift, we must separate the clocks in height; therefore, we must connect them by a signal that traverses the distance between them. But this makes it impossible to determine unambiguously whether the shift is due to the clocks or to the signal. The observable phenomenon is unambiguous: the received signal is blueshifted. To ask for more is to ask questions without observational meaning.

This is a key aspect of relativity, and in fact of physics as a whole. We concentrate only on quantities that we can *measure* with physical devices, and avoid unanswerable questions.

There is one way to see the effect of the gravitational redshift without an intervening signal, however, and that is to measure its effect on the *elapsed* time of two clocks. Begin with two clocks side by side, ticking at the same rate, and synchronized, so that at some chosen moment they read the same time and tick at the same rate. Take one clock slowly to the top of the tower and let it sit there for a while. Then, bring it back down slowly and compare it with the ground clock. While the rates at which they tick will once again be the same once the clocks are reunited, the tower clock will be ahead of the ground clock. The inference from this is that the tower clock ran faster while it was on the tower, but unless we connect the clocks by a light signal, we cannot see the difference in the ticking rate except after the fact, once we reunite them. This idea actually was the basis for a 1971 experiment using atomic clocks and jet aircraft, to be described shortly.

Early attempts to measure the gravitational redshift focused on light from the Sun. When an atom undergoes a transition from one electronic level to another, it emits light at a frequency or wavelength that is a characteristic of the atom. In the laboratory, the frequencies of these "spectral lines" can be measured with high accuracy. The same atom on the surface of the Sun will emit light whose frequency is redshifted as seen from Earth because, in a thought experiment using a tower sitting on the surface of the Sun and stretching all the way to the Earth, the atom would be at the bottom of the tower and the receiver on Earth would be at the top of the tower (Figure 2.2). In this example we can ignore the effects of Earth's gravity and of its orbital motion; these produce a correction of order 0.03 percent to the dominant effect due to the Sun. For a wavelength of 5,893 angstroms (an angstrom is ten billionths of a centimeter), corresponding to the bright-yellow emission line of the sodium atom, one of the most intense in the solar spectrum, the shift is 0.0125 angstroms toward longer wavelengths (lower frequencies), well within reach of standard measurement techniques.

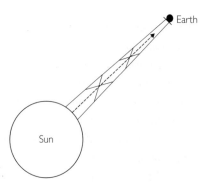

Fig 2.2 Gravitational redshift of light emitted by the Sun and received by an observer on Earth. The "tower" is used only to analyze the amount of the shift.

In 1917, however, Charles E. St. John of the Mount Wilson Observatory in California reported that he had found no "Einstein" redshift of spectral lines from the Sun, and a 1918 report from an observatory in Kodiakanal in India was inconclusive. One can only imagine how Einstein must have felt! Science historians think that these results had a direct negative impact on Einstein's candidacy for the 1918 Nobel Prize. The prize would not be awarded to him until 1921, and then only for his 1905 explanation of the photoelectric effect, and not for any of his relativistic theories. His theory of the photoelectric effect was verified definitively by experiment, whereas in 1921, verifying relativity still had a long way to go.

Looking back, we see the results of St. John and others not as a failure of Einstein's prediction but rather as a lack of understanding of the solar surface at the time. The gas at the surface of the Sun experiences violent and turbulent motions, with rising columns of hot gas and falling columns of cooler gas, which lead to Doppler shifts of the emitted frequencies both to the blue and to the red. The gas is also under high pressure, which causes shifts in the intrinsic frequencies emitted by certain atoms. These and other effects made it impossible in those early years to separate the gravitational shift clearly from other complex effects. It wasn't until the 1960s, when these effects were better understood, that astronomers were able to measure the gravitational redshift of solar lines. A measurement in 1991 confirmed the prediction to about 2 percent.

The problem with the Sun is that the relativistic shift is so tiny compared to other contaminating effects. But by 1920 astronomers had identified a few examples of a different kind of star, a white dwarf, which could be used to measure Einstein's predicted redshift. A white dwarf is a star with a mass comparable to that of the Sun, but compressed into a ball the size of the Earth, a hundred times smaller than the Sun. The gravitational redshift is thus about a hundred times larger than that from the Sun, and accordingly easier to detect. But the prediction of the redshift depends on the mass and radius of the white dwarf, which are not as well known as those quantities are for the Sun.

Fortunately, there was an exception to this even as far back as the 1920s. One of these unusual stars, called Sirius B, was actually in orbit around the "dog star," Sirius (called Sirius A), the brightest star in the night sky. This allowed astronomers to determine that Sirius B has about the same mass as the Sun, as inferred from its orbital motion around Sirius A. In 1924, Arthur Stanley Eddington (1882–1944), who, in addition to his talents as an astronomer, was the world's leading expert in stellar structure at the time, used his mathematical models to argue that the radius of Sirius B is about forty times smaller than that of the Sun. From that, he made a prediction of the gravitational redshift of spectral lines from Sirius B. At the same time, the noted spectroscopist Walter S. Adams of the Mount Wilson Observatory in California, who had first measured the spectrum of Sirius B in 1915, was engaged in making improved measurements with a view toward detecting the Einstein redshift. The spectrum was difficult to interpret, in part because of contamination of light from the much brighter Sirius A. But in 1925 Adams reported his results, in remarkably close agreement with Eddington's prediction. The *New York Times* reported "New Test Supports Einstein's Theory."

In time, however, it all began to unravel. First, it was realized that the models used by Eddington to study white dwarfs were wrong. His Cambridge University colleague Ralph Fowler pointed out in 1926 that the white dwarf was an entirely new kind of astronomical beast, with an internal constitution radically different from normal stars, governed by the quantum mechanical principle that no two electrons can occupy the same state. In addition, as more white dwarfs were discovered and their unique spectral signatures identified, Adams' interpretation of his spectra also came under heavy fire.

The world would have to wait another forty years before the white dwarf test could be carried out correctly. It was not until 1961 that the orbit of Sirius B would bring it far enough away from Sirius A as seen from Earth to enable new and improved spectral measurements, now using the Mount Palomar 200 inch telescope. Meanwhile, modern theories of white dwarf structure had been developed that could make better predictions of the radius and of the redshift. Finally, in 1971 results were announced that confirmed Einstein, but showed that the redshift inferred from the spectra was four times larger than that claimed by Adams, and the predicted shift inferred from the theoretical models was four times larger than that predicted by Eddington. Adams and Eddington were both off by the same factor of four! Suggestions have been made of conscious or unconscious bias on Adams' part, given that he and Eddington were in regular communication while he was making his measurements. But most science historians reject this claim, arguing that both the theory and the observations circa 1924 were so compromised that it was pure luck (good, bad or otherwise) that Adams and Eddington got apparent agreement. In 2005, measurements of Sirius B using the Hubble Space Telescope confirmed the Einstein redshift to about 6 percent.

Einstein had proposed three crucial tests of his theory: the explanation of an anomalous advance of the orbit Mercury, which he had worked out in his 1915 papers (page 119), the deflection of light, as confirmed in 1919, and the gravitational redshift. In 1950, Einstein had to admit that evidence for the redshift effect was "not yet confirmed." Within ten years, however, confirmation of the redshift would finally arrive, not from astronomy but from the physics laboratory.

The first truly accurate and reliable test of the redshift was the Pound–Rebka experiment of 1960. This experiment is very close in concept to the one described in our thought experiment in Figure 2.1. In this case, the tower was the Jefferson Tower of the physics building at Harvard University. For the tower's height of 74 feet the predicted frequency shift is only two parts in a thousand trillion, and thus an emitter and receiver of extremely well-defined frequency are required. Robert V. Pound and his student, Glen Rebka, Jr., used the unstable Fe^{57} isotope of iron, which has a lifetime or half-life of one ten-millionth of a second (one tenth of a microsecond). When this isotope decays it emits light in the form of gamma rays of wavelength 0.86 angstroms, within a very narrow range in wavelengths of only one part in a trillion of the basic wavelength. The same isotope can also absorb gamma rays of the same wavelength within the same narrow spread.

However, this alone is not enough to measure the redshift. Inside any realistic sample containing Fe^{57}, the iron nuclei are constantly in random motion because of the internal energy contained in any body at a finite temperature. This leads to Doppler shifts in the emitted gamma ray frequencies, whose result is to broaden or smear the range of wavelengths. In addition, upon emission or reception of a gamma ray, the iron nucleus "recoils," just as a billiard ball will recoil slightly if struck by a ping-pong ball, and this recoil velocity also causes a Doppler shift in the frequency. These effects can broaden the range of frequencies emitted and absorbed by a realistic Fe^{57} sample so severely that a redshift measurement would have been impossible, had it not been for Rudolph Mössbauer.

Working at the Max Planck Institute in Heidelberg, Germany, in the late 1950s, Mössbauer discovered that if an iron nucleus is implanted in the right kind of crystal, then the forces of the surrounding atoms not only reduce the heat-induced motions of the atom, but also transfer the recoil of the emitting atom to the crystal as a whole, thereby virtually eliminating it, because of the enormous mass of the crystal compared to the iron atom. For this discovery, Mössbauer was awarded the Nobel Prize in Physics in 1961. The Harvard gravitational redshift experiment was one of the many important applications of the effect cited at the awards ceremony by the Swedish Academy of Sciences, which is in charge of the Nobel Prize.

Pound and Rebka carefully fabricated their Fe^{57} emitters and absorbers in order to take best advantage of the Mössbauer effect. But the range of frequencies emitted and absorbed was still a thousand times larger than the size of the expected gravitational shift, so they used a clever trick.

They put the emitter on a movable platform that could be raised and lowered slowly using a hydraulic lift and a rack-and-pinion clock drive. If the emitter was at the top of the tower, so that the gamma rays would be blueshifted upon reaching the bottom, the platform was raised slowly, producing a Doppler shift toward the red. By adjusting the rate at which the emitter was raised, Pound and Rebka could produce a Doppler red-shift that would cancel or compensate for the gravitational blueshift, thereby allowing the range of frequencies of gamma rays received at the bottom to match closely the range that could be absorbed by the receiver. The Doppler shift required to do this was then a measure of the gravitational blueshift. The needed velocity was about 2 millimeters per hour. In order to eliminate certain sources of error, the actual experiment was a symmetrical one. Half the measurements were made with an emitter at the top and an absorber at the bottom, to measure the blueshift, and half were made with an emitter at the bottom and an absorber at the top, to measure the equal and opposite redshift. The results of the 1960 experiment agreed with the prediction to 10 percent, and those of an improved 1965 experiment version by Pound and Joseph L. Snider agreed to 1 percent.

As we mentioned earlier, another way to check the gravitational redshift is to compare the readings of two clocks that are separated temporarily. During October 1971, a remarkable experiment was performed that checked both these phenomena—gravitational redshift and special relativity's time dilation—in their effects on traveling clocks. The idea behind the "jet-lagged clocks" experiment is this. Consider, for simplicity's sake, a clock on Earth's equator, and an identical clock on a jet plane flying overhead to the east at some altitude. Because of the gravitational blueshift, the flying clock will tick faster than the ground clock. What about special relativity's time dilation? Here we must be a bit careful, because the Earth is also rotating about its axis, so both clocks are moving in circles around the center of the Earth, rather than in straight lines through empty space.

According to special relativity, the rate of a moving clock must always be compared to a set of clocks that are in an inertial frame, in other words that are at rest or moving in straight lines at constant velocity. Therefore we can't simply compare the flying clock directly with the ground clock. Let us instead compare the rates of both clocks to a set of fictitious clocks that are at rest with respect to the center of the Earth and not rotating with the Earth (Figure 2.3). The ground clock is moving at a speed determined by the rotation rate of the Earth, and thus ticks more slowly than the fictitious inertial clocks (as represented by a master inertial clock in Figure 2.3); when the flying clock is moving in the same

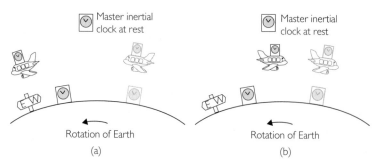

Fig 2.3 Jet-lagged clocks. (a) Eastbound. At the start, the flying clock is directly above the ground clock (in gray); after some time the ground clock has moved, while the flying clock is farther to the east. The flying clock has traveled more quickly than the ground clock relative to a stationary master clock, and therefore ticks more slowly relative to it than does the ground clock because of the time dilation of special relativity. Thus, the flying clock ticks more slowly than the ground clock. On the other hand, the gravitational blueshift makes the flying clock tick more quickly than the ground clock. The two effects can offset each other. (b) Westbound. At the start, the flying clock is again directly above the ground clock; after some time the ground clock has moved, but the flying clock has not moved as far because a typical commercial jet cannot overtake the Earth's rotation. It has traveled more slowly relative to the inertial clock than has the ground clock. Thus, the flying clock ticks more quickly than the ground clock relative to the master clock because of time dilation. The gravitational blueshift also causes the flying clock to tick more quickly, so the two effects add to each other.

direction as Earth's rotation (eastbound), then it is moving even more quickly than the ground clock relative to the inertial clocks, so it is ticking even more slowly. Thus time dilation makes the flying clock run slowly relative to the ground clock.

In this thought experiment, the two effects, gravitational blueshift and time dilation, tend to offset one another, and whether the net effect is that the flying clock ticks more quickly or ticks more slowly than the ground clock will depend on the height of the flight, which determines the amount of gravitational blueshift or speed-up, and the ground speed of the flight, which determines the amount of time dilation slow-down. Consider now a westbound flying clock at the same altitude. The gravitational blueshift is the same, but now the flying clock is traveling more slowly relative to the inertial clocks than is the ground clock, and therefore it is the ground clock that ticks more slowly relative to the inertial clocks. Therefore, the flying clock ticks more quickly than the ground clock, and in this case both the gravitational and time dilation effects work together, causing the flying clock to tick more quickly. If we were to start with three identical, synchronized clocks, and we were to leave one at home while sending one around the world to the east and the other around the world to the west, we would expect the westward clock to return having gained time, or aged more quickly, while the eastward clock would have gained or lost time depending on the altitude and speed of the flight.

The actual experiment, coordinated by J. C. Hafele, then at Washington University in St. Louis, and Richard Keating of the US Naval Observatory, used cesium-beam atomic clocks. Because of their limited budget they could not simply charter planes to circumnavigate the globe non-stop, but instead they had to fly the clocks on commercial aircraft during regularly scheduled flights. (Because of government regulations, they couldn't even fly first class!) No, the clocks were not strapped into their seats like the other passengers. Actually on most of the flights they were positioned against the front wall of the coach-class cabin to protect them against sudden motions on landing and to connect them more easily to the airplane's power supply. The flights included numerous stopovers during the course of the experiment, and the air speeds, altitudes, latitudes and flight directions all varied. But by keeping careful logs of the flight data, they could calculate the expected time differences for each flight. The eastward trip took place between 4 and 7 October and included 41 hours in flight, while the westward trip took place between 13 and 17 October, and included 49 hours in flight. For the westward flight the predicted gain in the flying clock was 275 nanoseconds (billionths of a second), of which two thirds was due to the gravitational blueshift; the observed gain was 273 nanoseconds. For the eastward flight, the time dilation was predicted to give a loss larger than the gain due to the gravitational blueshift, the net being a loss of 40 nanoseconds; the observed loss was 59 nanoseconds. Within the experimental errors of plus or minus 20 nanoseconds, attributed to inaccuracies in the flight data and intrinsic variations in the rates of the cesium clocks, the observations agreed with the predictions!

What is the main ingredient missing from the two experiments we have just described? Height. Up to a limiting value, the size of the gravitational redshift increases with the difference in height between the emitter and receiver or between the two clocks. The limit arises because the higher you go, the weaker gravity gets, so that eventually added height makes no difference. The gamma-ray experiment of Pound and Rebka, unfortunately, could not go to larger differences in height, because the gamma rays are emitted equally in all directions from the sample of Fe^{57} crystals. A consequence of this is that as the height is increased, the number of gamma rays received by the absorber becomes so small as to be unusable. The jet-lagged clocks experiment was financially limited to typical commercial aircraft altitudes. But what about putting an atomic clock on a satellite or rocket? By the time of the Hafele–Keating experiment, plans for such an experiment were already under way.

The idea of a satellite test of the redshift had been suggested as early as 1956, just before the first Earth satellites were launched, and had been tried in 1966, with modest success, at the 10 percent level. But the experiment that was being worked on in 1971 was truly ambitious. The idea was to get two of the best atomic clocks in existence, called hydrogen maser clocks, put one on the top of a rocket, blast it up to a couple of times the radius of the Earth, and compare its rate to the clock left on the ground as the rocket ascends and later descends back to Earth. In principle, the accuracy achievable in a measurement of the shift was one part in ten thousand, or one hundredth of a percent. Fortunately, the experiment brought together the two sets of experts required to pull it off.

The first set of experts consisted of Robert Vessot and Martin Levine of the Smithsonian Astrophysical Observatory at Harvard University. Their laboratory was at the forefront of development of this new kind of atomic clock. Soon after the invention of the hydrogen maser clock in 1959 by Harvard physicists Norman Ramsey, Daniel Kleppner and H. Mark Goldenberg, Vessot, who then worked for Varian Associates, pioneered the development of a commercial, portable version of the new timepiece. By 1969 Vessot had left industry for Harvard, and now wanted to make use of these devices in fundamental physics experiments. The other set of experts was the National Aeronautics and Space Administration

(NASA), which would provide the launch vehicle, tracking and other facilities required to get the clock aloft and measure its frequency shift.

The hydrogen maser clock was ideal for this experiment. It is based on a transition between two atomic energy levels in hydrogen that emits light in the radio portion of the spectrum, with a frequency of 1,420 million cycles per second (1,420 megahertz), or a wavelength of about 21 centimeters. The spread in frequencies is so narrow that the actual frequency is known to twelve significant digits, or to an accuracy of one part in a hundred billion.

The original plan was to put one of these clocks in orbit, but by 1970 it was evident that the cost of the Titan 3C rocket and the 2,000-pound payload required to achieve this was more than the NASA budget would permit, and a more modest plan was developed. The new plan was to send the clock on a suborbital flight to an altitude of about 10,000 kilometers, a thousand times higher than typical commercial jet altitudes, using a cheaper Scout D rocket and a smaller payload of a few hundred pounds. NASA denoted the project Gravity Probe-A. But there were two problems that had to be overcome to make the experiment work.

The first was to build a lightweight maser clock that would withstand the 20 g acceleration it would experience during launch. The second was how to detect the gravitational redshift. Consider what happens during the ascent of the rocket, say, when the rocket clock emits its signal, and the signal is received at the ground and compared with the frequency of the ground clock. The received frequency differs from the ground clock frequency because of two effects: the gravitational blueshift, caused by the height difference, and special relativity's time dilation, caused by the rocket's rapid motion. However, the received frequency is also shifted toward the red because of the usual Doppler shift produced by the rocket's motion away from the ground clock (during descent of the rocket, the Doppler effect would produce a blueshift), and this Doppler shift is 100,000 times larger than the gravitational redshift for a typical Scout D velocity of several kilometers per second.

One clearly must find a way to eliminate this huge effect somehow, in order to see the much smaller effects of interest, and this was done in a very elegant way as follows. Suppose a signal is emitted from the ground clock toward the ascending rocket clock (this is called the "uplink," while a downward signal is called the "downlink"). Incorporated into the rocket payload is a "transponder," a device that takes a received signal and sends it right back with the same frequency (and with a little more power, to make up for any losses during transmission up). Right when the transponder sends the transponded downlink, a second "one-way" downlink is sent to the receiver on Earth (see Figure 2.4).

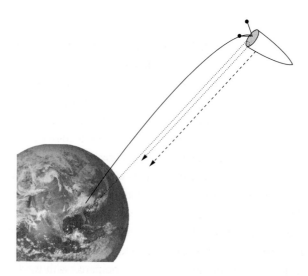

Fig 2.4 Gravity Probe-A. As a hydrogen maser clock rises over the Atlantic in the nose cone of a Scout rocket, a signal from an identical ground maser clock is sent toward it. When the signal is received by the rocket, it is sent back, and a signal directly from the rocket clock is sent along with it. Because the rocket clock is at a different height and is moving at a different velocity than the ground clock, the frequency of the one-way signal received by the ground clock is changed by the Doppler effect, the gravitational redshift, and special relativity's time dilation. For the two-way signal, though, the ground clock is both emitter and receiver, so there is no gravitational redshift or time dilation (the signal is emitted and received at the same height and velocity), and the Doppler shift contributes twice. The transponder that turns the signal sees a signal redshifted by the Doppler effect because it is moving away from the ground, and the ground receiver sees the signal Doppler shifted again because the transponder that turned it around is moving away from the ground. Thus, if half of the frequency change of the round-trip signal is subtracted from the frequency change of the one-way signal, the Doppler effect will cancel out completely.

What happens to the frequencies of the transponded signal and that of the one-way downlink signal when received back on Earth? Initially, when the uplink signal arrives at the ascending rocket clock, the received frequency differs from that of the rocket clock by the Doppler shift, and by the gravitational redshift and time dilation. Then, when the transponded signal is received back on Earth, its frequency is further redshifted by the Doppler effect, because the transponder (which you recall is attached to the ascending rocket) is receding from Earth, but it is now gravitationally blueshifted by an amount that exactly cancels the gravitational redshift experienced by the signal on the uplink. The signal is also changed by time dilation, but also in a way that exactly cancels the change experienced on the uplink. Therefore, when received back at the ground, this two-way signal has had its frequency changed by *exactly* twice the Doppler shift, and that's all. On the other hand, the one-way downlink signal sent by the ascending rocket clock and received at the ground has been changed by only one factor of the Doppler shift and by the gravitational blueshift and the time dilation. All one has to do then is take the frequency change on the two-way signal, divide by two, and subtract it from the one-way frequency change, and presto: no Doppler effect. This Doppler cancelation scheme was in fact incorporated directly into the electronics that gathered the data from the two radio links, and so it disappeared from the experiment altogether.

After the years of development of the clocks, of making one of them spaceworthy, of testing and retesting them to simulate launch conditions, the time had come to actually do the experiment. It was a perfect day for a launch, a pleasant June morning in 1976, with just some high, thin clouds in the sky. Vessot was in charge of the rocket clock at NASA's launch facility on Wallops Island, one of the many small islands that hug the eastern coast of the narrow Virginia peninsula separating Chesapeake Bay from the Atlantic Ocean. Levine took care of the ground clock at the NASA tracking station at Merritt Island, right next to Cape Canaveral in Florida. As is often the case, the period leading up to launch was not without its crises. One countdown had already been aborted because of a problem with some ammonia refrigerant. A misbehaving monitor designed to keep track of conditions in the rocket clock was brought into line by Vessot through the elegant technique of dropping it on the floor, a sort-of old-fashioned "hard-reboot."

Finally, the countdown reached the end, and at 6:41 a.m. Eastern Standard Time the Scout D roared into the Virginia skies. At 6:46, the payload containing the clock separated from the fourth stage of the rocket, and was in free fall thereafter. At this point data could be taken, because the rocket clock was no longer affected by the high accelerations and vibrations of launch. For about three minutes the one-way downlink frequency from the rocket clock (with the Doppler piece canceled automatically, remember) was lower than that of the ground clock, because the high velocity of the rocket caused a time dilation redshift to lower frequencies, while the altitude was not yet large enough to produce a gravitational blueshift. At 6:49 the frequencies of rocket and ground clock were exactly the same, because the gravitational blueshift canceled exactly the time dilation redshift. After that, as the altitude increased and the speed of the rocket decreased, the gravitational blueshift dominated more and more. The peak of the orbit occurred at 7:40. Here, the shift was predominantly the gravitational blueshift, amounting to almost 1 hertz out of 1,420 megahertz, or four parts in ten billion. Because both the rocket clock (after separation from the fourth stage of the Scout D) and the ground clock maintained their intrinsic frequencies stably to one part in a million billion, they could measure these changes in frequency to very high accuracy. Data taking continued during descent, with the cancelation between gravitational blueshift and time dilation occurring again at 8:31. At 8:36, the payload was too low in the sky to be tracked reliably, and shortly thereafter, some 900 miles east of Bermuda, the rocket and its onboard atomic clock crashed as planned into the Atlantic Ocean. This two-hour flight produced more than two years of data analysis for Vessot and his colleagues, but when all was said and done, the predicted frequency shifts agreed with the observed shifts to a precision of seventy parts per million, or to 7/1000 of a percent.

Modern atomic clocks maintain time so precisely that the gravitational redshift now touches our daily lives. This remarkable convergence between fundamental physics and everyday life is due to GPS, the Global Positioning System. Deployed originally for military navigation, GPS has rapidly transformed itself into a thriving commercial entity with countless applications. It is based on an array of as many as thirty-two satellites orbiting the Earth, each carrying a precise atomic clock based either on cesium or rubidium atoms. Using a GPS-enabled device, which detects radio emissions from any of the satellites which happen to be overhead, users can determine their absolute latitude, longitude and altitude to an accuracy of 15 meters, and local time to fifty billionths of a second. In addition to GPS there are the Russian GLONASS, the Chinese BeiDou and the European Galileo systems, all in various stages of operation and development.

Apart from the obvious military uses, GPS has found applications in airplane navigation, oil exploration, wilderness recreation, bridge construction, sailing and interstate trucking, to name just a few. Another crucial application is finding lost smartphones and lost pets! If you purchased this book from a vendor such as Amazon, its route to your door was tracked every step of the way using barcode scanners with GPS. Even Hollywood has met GPS, pitting James Bond in the 1997 film *Tomorrow Never Dies* against an evil genius who was inserting deliberate errors into the GPS system and sending British ships into harm's way. For better or for worse, GPS is everywhere and it is here to stay.

But how does GPS actually work? On a piece of paper in two dimensions, the problem is one faced by every high school mathematics student: given two fixed points representing two satellites, find the point, representing your cell phone, that is a given distance from one and a given distance from the other (Figure 2.5). The problem is solved by drawing arcs of the given radii using a compass and finding where they intersect. Your cell phone does something similar, albeit a bit more complicated. First, the GPS receiver in your cell phone calculates the distance between itself and each satellite with which it communicated, using the time at which the signal was emitted, as determined by the on-board atomic clock and encoded into the signal, the time at which the signal was received by your phone, and using the speed of light. Then, given readings from four GPS satellites, it is a simple matter to use the same principle as our high school math student to compute the receiver's precise location, both in space and in time (the GPS receiver does it by

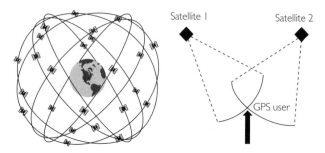

Fig 2.5 GPS and relativity. As many as thirty-two satellites of the US GPS system orbit the Earth in an array of orbital planes designed so that a user has a strong likelihood of connecting to three or four satellites at any given time. Knowing the distances from any two satellites on a plane, a user can determine his location on that plane by the analogue of finding where two circles of the given radii intersect on the plane, as shown in the right panel. With three satellites in view, the user can determine his location in three dimensions. With four satellites, the user can also determine local time, i.e. the fourth dimension.

solving equations embedded in the computer chip, not with compasses). To achieve a navigation accuracy of 15 meters, time throughout the GPS system must be known to an accuracy of 50 nanoseconds, which simply corresponds to the time required for light to travel 15 meters.

But, the orbiting clocks are 20,000 kilometers above the Earth in accurately known orbits that circle twice per day, and experience gravity that is four times weaker than that on the ground. Because of the gravitational redshift effect, the orbiting clocks tick slightly faster, by about 45 microseconds (millionths of a second) per day, than ground clocks. The satellite clocks are also moving at 14,000 kilometers per hour, much faster than clocks on the surface of the Earth, and special relativity says that such rapidly moving clocks tick more slowly, by about 7 microseconds per day, because of time dilation. The net result is that time on a GPS satellite clock advances faster than a clock on the ground by about 38 microseconds per day. Compare that with the 50 nanoseconds precision required! At 38 microseconds per day, the relativistic offset in the rates of the satellite clocks is so large that, if left uncompensated, it would cause navigational errors that accumulate faster than about 7 meters per minute!

When the first GPS satellite was launched in 1977, it was already recognized that incorporating relativity would be necessary. Initially this was done by electronically adjusting the rates of the satellite clocks so that they artifically ticked at the same rate as ground clocks.

But in 1983, after some relativity experts argued that the relativistic effects were being implemented incorrectly, the US Air Force, which ran the GPS program at the time, began to worry. Six satellites were already in orbit with more scheduled for launch, and it became important to determine if these critics were correct. Accordingly, they asked the Air Force Studies Board (AFSB), a part of the National Research Council, to conduct an independent analysis of their methods. That board then asked Cliff (Nico was only three years old at the time!) to put together a committee of experts and to chair the study. After examining the methods used by the Air Force to account for relativistic effects, and after studying the analysis of the critics, Cliff's committee concluded that the criticisms were not correct and that the Air Force was properly implementing methods that had become standard in the atomic clock community. The committee then turned to a list of tasks of a more operational nature for which the Air Force wanted advice.

An awkward moment occurred midway through the year-long study when the AFSB staff person asked Cliff when he had become a naturalized US citizen. Cliff replied that he was still a Canadian citizen on a permanent resident's green card (a fact evidently overlooked on the curriculum vitae that he had submitted to the AFSB). This was a problem, because by AFSB rules all studies are initially classified, even if they do not actually deal with classified material, as was true of Cliff's study. After some urgent phone calls the staffer managed to get a general to declassify the study retroactively, so that the work could continue.

At the end of the study Cliff was required to make a final report to a meeting of the AFSB, headed by a four-star general who was at the time in charge of the Air Force Systems Command. Confronted with the challenge of explaining general relativity and its importance for GPS to an audience more familiar with military matters than with Einstein's theories, Cliff prepared the most simple and colorful viewgraphs possible (this was in the pre-PowerPoint era), and tried to make his presentation lively and engaging. Unfortunately, within thirty seconds the general's eyelids started closing…closing…until eventually he fell asleep! Luckily his aide, a second lieutenant, took careful notes, and Cliff presumes that she briefed the General afterward on what he said. But still, for a general relativist to be briefing a military leader on Einstein's theory because it was important for US national security, it didn't matter whether he was awake or asleep, this was a moment to be remembered!

Without the proper application of relativity GPS would fail in its navigational functions within about two minutes. So the next time your plane approaches an airport in bad weather, and you just happen to be wondering "What good is Einstein's theory?", think about the GPS tracker in the cockpit, helping the pilots guide you to a safe landing.

Let us return to using clocks to test Einstein's equivalence principle. Over the 40 years since Vessot's rocket experiment using hydrogen maser clocks, atomic timepieces have become much better. During the 1980s a new technique was discovered that allowed researchers to use crossed laser beams tuned to specific wavelengths to trap and slow down clouds of atoms. One of the leading enemies of time precision is the Doppler effect caused by the random motions of atoms, which smears the fundamental frequency of the light emitted by the atom, making it slightly less precise as a standard for measuring time. But the new method of "laser cooling" could slow down atoms to such a degree that the only way to characterize how fast they were moving was to express the speed in terms of the apparent temperature of a gas, with absolute zero (0 kelvin or $-459.67°F$) representing absolute stillness. Temperatures measured in millionths of a degree above absolute zero (microkelvins) are now achieved routinely. In Vessot's clocks the hydrogen atoms were at room temperature, about $300°K$ warmer.

New ideas for ever more accurate and stable clocks have been developed, based on such concepts as "atom fountains," "Bose–Einstein condensates" and "atom interferometry." To describe these would take us too far afield, but we cannot resist mentioning one experiment that demonstrates just how "everyday" Einstein's gravitational redshift has become.

This was an experiment done in 2010 in the laboratory of David Wineland at the National Institute of Standards and Technology in Boulder, Colorado. The experimenters set up two clocks based on ultracold trapped aluminum ions, separated by a height of only 33 centimeters, or about a foot. They were able to measure that the higher clock was ticking a bit faster than the lower clock. If you have ever wondered why it seems that your brain is aging faster than your feet, now you have the answer. But cheer up, the difference is only about 7 nanoseconds per year.

The effects we have been discussing, whether in the laboratory, on GPS satellites or on white dwarfs, are extremely tiny differences in the rate at which time moves forward. So what were the crew of the *Endurance* talking about: one hour equaling seven years back on Earth? How can that be possible? The answer is that the warpage of time becomes that extreme near the event horizon of a black hole. In *Interstellar*, Miller's planet was orbiting very close to Gargantua's horizon. As a result, during their stay on Miller's planet their clocks would be slower than Earth's clocks by a whopping factor of 60,000. In fact, the closer you get to the horizon of a black hole, the larger this factor becomes. Upon Cooper's return to Earth (spoiler alert!) the effect of this warpage of time leads to a tear-inspiring scene as he visits his daughter Murph, now a very old woman.

These numbers were not invented out of thin air by the movie's director Christopher Nolan and the screenwriter, his brother Jonathan. In fact, many of the details related to the black hole Gargantua and the planets revolving around it were worked out carefully using Einstein's theory by Caltech astrophysicist Kip Thorne, who developed the original concept on which the movie was based and was an executive producer of the film. In addition to his talents as a moviemaker, Christopher Nolan is a self-confessed "science geek," and so he wanted Thorne to help make the movie as scientifically accurate as possible, within the confines of the science fiction genre. For example, the plot of the movie requires that the crew pass through a "wormhole," which is a popular science fiction tool, but which no relativist, including Thorne, believes is possible in nature according to our current understanding of the laws of physics.

But, is the warpage of time near a black hole a purely theoretical thing, useful for nothing more than movie plot twists? Not at all, because on 19 May 2018 a star designated by the prosaic name S2 passed its point of closest approach to the supermassive black hole known as Sgr A* that resides in the center of our Milky Way galaxy, a black hole with the mass of 4.3 million Suns. During that close encounter, reaching just 120 times the Earth–Sun distance away from the horizon, astronomers using advanced infrared telescopes in Chile and Hawaii were able to measure Einstein's gravitational redshift of the spectrum of S2. But you'll have to wait until Chapter 6 to read more about this, when we discuss how to test Einstein's theory near black holes.

CHAPTER 3

How Light Sheds Light on Gravity

After over a year of painstaking preparation and numerous rehearsals, Don Bruns felt he was ready. The skies over Casper, Wyoming that morning were clear and blue, with only a few thin clouds. Winds were calm. His TeleVue Optics NP101is telescope and its attached CCD camera were ready. The computer programs he had written to send commands to the telescope during the crucial two and a half minutes had been tested and rehearsed. All that remained was to sit back and wait for what was being called the Great American Eclipse.

Bruns was one of an estimated 215 million people who viewed the eclipse of the Sun either in person or electronically that 21 August 2017. But Bruns, a retired physicist and amateur astronomer, was not in Wyoming to ooh and ahh over the sight of the eclipse, he was there hoping to repeat one of the most famous measurements of the twentieth century, an experiment that made Albert Einstein famous to the world at large.

That earlier experiment generated over-the-top headlines in the autumn of 1919. "Revolution in Science / New Theory of the Universe / Newtonian Ideas Overthrown," proclaimed *The Times* of London on 7 November. "Lights all Askew in the Heavens / Men of Science More or Less Agog over Results of Eclipse Observations," declared the *New York Times* three days later (notice only "men" of science, a typical attitude of that era). It heralded a brave new world in which the old values of absolute space and absolute time were lost forever. To a world emerging from the devastation of World War I, it meant the overthrow of all absolute standards, whether in morality or philosophy, music or art. In a 1983 survey of twentieth-century history, the British historian Paul Johnson argued that the "modern era" began not in 1900 or in August 1914, but with the event that spawned these headlines in 1919.

This event made Einstein a celebrity. Set aside for a moment his genius, the triumph of his theories, and the new scientific order he created almost singlehandedly. That alone might have been enough, but Einstein was also, in today's terminology, a very "media-friendly" fellow. His absentmindedness, his wit, his willingness to expound upon politics, religion and philosophy in addition to science, his violin playing—all these characteristics sparked an intense curiosity on the part of the public. The press, tired of printing battle reports and casualty lists from the war, was only too eager to satisfy its readers' curiosity.

The event that caused such a commotion was the successful measurement of the bending of starlight by the Sun. The amount of bending agreed with the prediction of Einstein's general theory of relativity, but disagreed with the prediction of Newton's gravitational theory. This was the experiment that Don Bruns planned to repeat, and if all went well, to beat in precision.

The story of how gravity affects the trajectory of light is one of the most fascinating in all science. It actually has its roots in the eighteenth century, yet the story continues to evolve to this day. It journeys from the heights of theoretical and experimental accomplishments to the depths of racist propaganda, from our solar system to the most distant galaxies.

It is believed that the first person to consider seriously the possible effect of gravity on light was a British theologian, geophysicist and astronomer, Reverend John Michell (1724–1793). Ever since the time of Newton, who had himself speculated vaguely that gravity might affect light, it had been assumed that light consisted of particles or "corpuscles." In 1783, Michell reasoned that light would be attracted by gravity in the same way that ordinary matter is attracted. He noted that light emitted outward from the surface of a body such as the Earth or the

Sun would slow down after traveling great distances (Michell, of course, did not know the theory of special relativity, which requires the speed of light to be the same from the viewpoint of any inertial observer). He then asked how large would a body of the same density (same number of grams per cubic centimeter) as the Sun have to be in order that light emitted from it would be stopped by gravity and pulled back before escaping? The answer he obtained was 500 times the diameter of the Sun. Light could never escape from such a body.

This remarkable idea describes what we now refer to as a black hole. In today's language, Michell's object would be 100 million times more massive than our Sun. Fifteen years later, the great French mathematician Pierre Simon Laplace performed a similar calculation. Although Michell and Laplace were wrong in the fundamental theory, their basic premise is right: gravity affects light.

But Michell didn't stop there. He then asked, how would one ever detect such a body if light could not escape from it? His remarkable answer was that if such a dark body were to be in a double star or binary orbit with a normal star, one could infer its existence by measuring the wobble in the normal star's position as the two bodies revolved around each other. What made this remarkable was that in 1783 there was no evidence that such binary star systems existed. It appears that Michell's writings and speculations on such possibilities, along with his groundbreaking statistical analyses of close associations of stars on the sky, played a role in getting astronomers to start looking for evidence of binaries. The first solid discovery of two regular stars in a mutual orbit was announced by Wilhelm Herschel in 1803.

Michell's friend and colleague, Henry Cavendish (1731–1810), shared his interest in gravity. Already famous for his discovery of hydrogen, Cavendish inherited from his recently deceased friend an instrument that Michell had built to measure gravity, and after some modifications of it, he used it to measure what we now call Newton's constant of gravitation (called "big G" by physicists), which relates the gravitational force between two bodies to their masses and separation.

But around 1784, Cavendish also asked the question: If gravity affects light as Michell suggested, would it not also bend it? According to

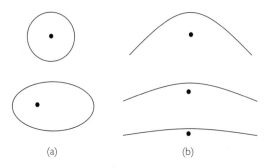

Fig 3.1 Newtonian orbits. Orbits (a) are bound, either circular or elliptical (also called eccentric). Orbits (b) are unbound hyperbolae of ever greater speed, moving from top to bottom.

Newtonian gravity, the orbit of one body about another is a "conic section," the figure formed by the intersection of a cone with a plane tilted at various angles: an ellipse or a circle if the orbit is bound, so that the body never escapes, or a hyperbola if it is unbound (Figure 3.1). If light is a corpuscle undergoing the same gravitational attraction as a material particle, then because its speed is so large, its orbit will be a hyperbola that is very close to being a straight line [the bottom panel of Figure 3.1(b)]. However, the deviation, while small, is calculable, and apparently Cavendish did the calculation.

Why apparently? Cavendish was notorious for not bothering to publish his work, or even to discuss it with colleagues (the neurologist Oliver Sacks has speculated that Cavendish may have had Asperger syndrome). Around 1920, during a project to complete the publication of Cavendish's work in physics (his work on chemistry having been compiled and published earlier), researchers discovered a scrap of paper among his documents which stated "To find the bending of a ray of light which passes near the surface of any body by the attraction of the body . . . ," followed by a formula. No calculational details, just the correct answer to the problem posed.

Some fifteen years after Michell and Cavendish, a similar story played out on the other side of the European continent, but with a somewhat different outcome. Prompted by Laplace's speculations, a Bavarian astronomer named Johann Georg von Soldner (1776–1833) asked the same question: Would gravity bend light? Von Soldner was a largely self-taught man who became a highly respected astronomer. He made fundamental contributions to the field of precision astronomical measurements known as astrometry, and eventually rose to the position of director of the observatory of the Munich Academy of Sciences. But in 1801, he was still an assistant to the astronomer Johann Bode in the Berlin Observatory. Von Soldner calculated the bending (his and Cavendish's answers agree), and determined that, for a path that skims the surface of the Sun, the bending would be 0.875 seconds of arc. An arcsecond is the angle subtended by a human finger at a distance of about 4 kilometers or 2.5 miles (3,600 arcseconds equals 1 degree of arc).

Von Soldner's work was published in 1804 in one of the German astronomical journals. It was then immediately forgotten, partly because the effect was beyond the current limits of telescope precision, and partly because of the rise during most of the nineteenth century of the wave theory of light, according to which light moves as a wave through an imponderable "aether," and presumably suffers no deflection. Einstein was certainly not aware of either von Soldner's paper or Cavendish's calculation. It was not until 1921 that von Soldner's work was rediscovered and resurrected, but then it was for a different, more unsavory purpose.

Like Cavendish and von Soldner over a century before, Einstein in 1907 was also interested in the effect of gravity on light. He recognized that if the principle of equivalence led to an effect on the frequency of light, the gravitational redshift (Chapter 2), it should also result in an effect on its trajectory. In 1911 he determined that the deflection of a ray grazing the Sun should be 0.875 arcseconds. He proposed that astronomers should look for the effect during a total solar eclipse, when stars near the Sun would be visible and any bending of their rays could be detected through the displacement of the stars from their normal positions (Figure 3.2). Several teams, including one headed by Erwin Finlay-Freundlich of the Berlin Observatory, one headed by William Campbell of the Lick Observatory in the USA, and one headed by Charles Perrine of the National Argentinian Observatory, traveled to the Crimea to observe the eclipse of 21 August 1914. But World War I intervened, and Russia sent many of the astronomers home, interned

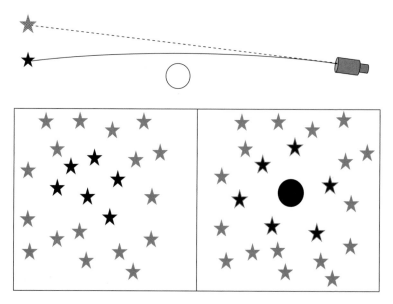

Fig 3.2 Deflection of light by the Sun. Top: Deflection of light causes the apparent position of a star to be displaced away from the Sun. Bottom left: A field of stars viewed at night. Bottom right: The same field with the Sun in the middle, obscured by the Moon. Stars whose light passes close to the Sun (the six stars in black) have their locations displaced more than stars farther from the Sun (the stars in gray). The amount of bending is greatly exaggerated, of course.

others, and temporarily confiscated much of the equipment; in any case, the weather at the site on eclipse day would have been too bad to permit useful observations.

It is quite easy to see how Einstein's equivalence principle leads to a deflection of light. Imagine a laboratory with glass sides containing an observer well versed in the equivalence principle (see Figure 3.3). The laboratory is moving with constant speed far from any star or galaxy; it is therefore an inertial reference frame in which special relativity is valid. Because there is no gravity, the observer inside floats freely. The following sequence of events is shown in the upper panel of Figure 3.3: (a) A light ray enters the laboratory from the left at a spot right at the middle of the lab. (b) As the ray crosses the lab in a straight line, the lab moves forward (upwards in the figure), so the ray is still at the midpoint. (c) The ray exits the lab also at the midpoint. The ray crosses the lab in a straight line.

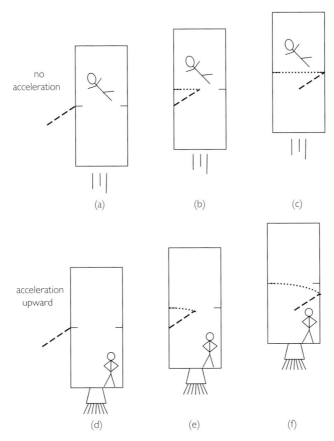

Fig 3.3 Einstein's principle of equivalence and the bending of light. Top panel: A light ray (dashed lines) enters a laboratory moving at constant speed in empty space. It crosses the laboratory in a straight line, with the angle changed because of aberration. Bottom panel: A light ray enters a laboratory accelerating via the thrust of a rocket. Because the laboratory is moving faster in each panel, the light ray leaves the laboratory slightly below the level where it entered, as if it were deflected downward.

But the angle of the ray as seen by the observer in the lab is different from the angle seen by external observers at rest! This is the well-known phenomenon of aberration, discovered by James Bradley in 1725. It manifests itself in an annual back and forth motion of stellar images by about 40 arcseconds as the Earth moves around the Sun. On a more mundane level, it is the same phenomenon that occurs when you carry an umbrella quickly through a vertical rainfall. You observe that the drops are tilted toward you and get your feet wet. In the top panel of Figure 3.3, the difference in angle between what we outside observers see for the light ray entering the lab and what the observer in the moving lab sees is substantial because we have made our lab move at around half the speed of light.

Now consider the same laboratory in distant space being accelerated by a rocket attached to it (the bottom panel of Figure 3.3). The observer can now stand on the floor of the lab because of the rocket's thrust. Let us assume that, when the light ray enters from the left, the speed of the laboratory is the same as it was in the top panel. The sequence of events now is different: (d) As the ray enters, it seems to be traveling once again horizontally as it initially enters the laboratory, because of aberration. (e) By the time the ray is half way across, the lab has traveled a bit farther than before because its speed is now a bit higher, so the ray is a bit below the lab's midpoint. (f) By the time the ray exits the lab, the lab has moved even more than before, and so the exit point is well below the midpoint. According to external observers, the light has traveled on a perfectly straight line (the dashed line), but according to the observer in the accelerating lab, the light ray (the dotted line) appears to have bent slightly toward the floor as it crossed the lab.

But, according to Einstein's principle of equivalence, the accelerating lab in the bottom panels of Figure 3.3 is equivalent to a lab at rest in a gravitational field. Therefore light should be bent by gravity! By considering a sequence of such laboratories all along the trajectory of a light ray passing by the Sun, and adding up all the tiny deflections, Einstein could conclude that the net deflection of a ray that just grazes the Sun would be 0.875 arcseconds. Therefore, whether we use the Newtonian theory of gravity combined with the corpuscular theory of light, as Cavendish and von Soldner did, or the principle of equivalence, as in this derivation, we predict the same deflection of light.

Yet, in November 1915 Einstein doubled the prediction. By that time, he had completed the full general theory of relativity, and found that, in a first approximation to the equations of the theory, the deflection had to be 1.75 arcseconds, not 0.875 arcseconds.

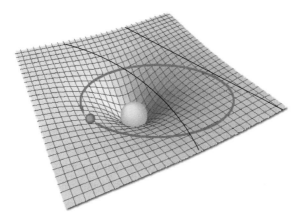

Fig 3.4 Rubber sheet analogy for curved space. A massive ball sits in the center of a rubber sheet stretched taut across a room. A small ball rolls around the banked surface, metaphorically representing a body in orbit around the large ball. Two fast bodies, metaphorically representing light rays, cross the sheet. The ray passing close to the ball is deflected because of the banking or warping of the surface.

Was this doubling completely arbitrary? Were the previous calculations wrong? Not at all. They are correct as far as they go. They simply did not, indeed could not, take into account an important circumstance that only the complete general theory of relativity could cope with: the curvature of space. As we saw in Chapter 2, the principle of equivalence tells us that time must be warped, but it says nothing about space. It is natural to assume that space might also be warped, since relativity involves uniting space and time into a unified spacetime. But it turns out that to determine by *how much* space is curved, we need the complete equations of general relativity, not just the principle of equivalence.

Indeed, general relativity predicts that space is curved near gravitating bodies, the curvature being greater the closer one gets to the body, and negligibly small at large distances. This is all embodied in the complicated mathematics of a four-dimensional spacetime continuum, and is very difficult to describe in words. However, if we strip away some unnecessary details, it is possible to present a *qualitative* picture that might give you a sense of some of the effects of curved space.

Let's imagine what curved space might look like around an object such as the Sun. Since we imagine the Sun is unchanging in time, we can strip away the time dimension and focus just on space. Also, since the Sun is spherical to a good approximation, like a soccer ball or basketball, any radial direction in space from its center is as good as any other, so we can pick one and then focus on the two-dimensional plane perpendicular to that direction. If we now ask what does that curved two-dimensional plane look like, it turns out that a good analogy is to imagine a rubber sheet stretched taut across a room on Earth, with a heavy bowling ball, representing the Sun, in the center (Figure 3.4). Because of the weight of the ball, it sinks and stretches the sheet. At the edges of the sheet, far from the ball, the sheet is approximately flat, with a geometry that obeys the usual rules of Euclid, but close to the ball it is warped.

As a result, the distance measured *along the rubber sheet* from the center of the indentation to the edge of the sheet is longer than the distance measured "as the crow flies," because of the stretching and warping of the sheet. The circumference of a circle drawn on the sheet around the ball would be smaller than 2π times the distance from the center measured along the sheet. In Euclidean geometry, the circumference of a circle is exactly 2π times the radius (with π a constant that is approximately $3.14159265\ldots$). A word of caution, however: this rubber sheet picture is only an analogy, and an imperfect one at that. Spacetime is not a rubber sheet; it is four-dimensional, not two, and it includes time as well as space.

Einstein's principle of equivalence says that, when viewed from a local freely falling frame, bodies move on straight lines, as if there were no gravity. But this is only a local statement, confined to a small region near the observer. Einstein argued that in a more general situation, in the presence of gravity, a body would move on the "straightest" line possible within the curved spacetime in which it found itself.

Such "straightest" lines are familiar to international air travelers. On the surface of the Earth, which is a true curved two-dimensional surface, such "straightest" lines are called geodesics. The equator is a geodesic, as are all the lines of longitude. The curve produced by the intersection of the Earth's surface with any plane passing through the Earth's center is also a geodesic. It is the path you would take if you traveled along the Earth's surface veering neither left nor right, but keeping the same forward heading. Your path is not straight in the Euclidean sense, because you are going around the Earth, ultimately returning to your starting point, but it is as straight as it can be. On Earth, such geodesic paths also happen to provide the shortest distance between two points, which is why your air trip from Los Angeles to Paris goes way north over Hudson's Bay and Greenland, and not over the mid-Atlantic. With the equivalence principle, Einstein extended this idea, asserting that freely moving particles travel on geodesics of a curved *spacetime*.

Let's now try to imagine how this idea about geodesics affects the path particles follow. In Figure 3.4, the small ball is orbiting the large ball by merely following the steeply banked rubber surface in as straight a manner as possible. This is just as in a roller coaster that turns to the right because the tracks are steeply banked in that direction, and not because it actually veers off the tracks (hopefully!). In a similar manner, a very fast particle, such as light, passes across the sheet near the edge with very little deflection (the nearly straight black line in Figure 3.4). But if such a particle passes close to the ball, it must dip down into the sheet, and because of the steep banking it will be deflected more.

The orbits and paths depicted in Figure 3.4 should actually be drawn using three dimensions, with time plotted somehow in addition, but that would severely challenge our artistic capabilities. Nevertheless, we hope that the rubber sheet metaphor, however imperfect, helps readers to visualize some of the consequences of curved spacetime.

The curvature of space explains Einstein's doubling. The previous calculations, such as the one using the accelerating laboratory or the one using Newtonian gravity, gave the deflection of light relative to space. If we thought that space was flat, that would be it. However, general relativity predicts that space is warped or "banked" near the Sun relative to space far from the Sun, and this adds an additional 0.875 arcseconds for a ray that grazes the Sun. Thus, the total deflection must be the sum of these two effects, or 1.75 arcseconds.

This space curvature effect is the important difference in the predictions of different theories of gravity. Any theory of gravity that is compatible with the equivalence principle (and almost all current theories are) predicts the first 0.875 arcseconds part. The second part comes from the curvature of space. Newtonian theory is a flat-space theory, so there is no further effect; the prediction remains at 0.875 arcseconds. General relativity, purely by coincidence, predicts an amount of space curvature that just doubles the deflection. Some theories predict slightly less curvature than general relativity, resulting in a slightly smaller value for the second part and a slightly smaller total deflection. Other theories predict more curvature, and thus a larger deflection angle.

Einstein's doubling of the predicted deflection had important consequences, for it meant that the effect was now a bit easier to observe. But the fact that a successful observation came as early as 1919, only four years after the publication of the general theory, must be credited to the pivotal role played by Arthur Stanley Eddington. We already encountered Eddington in Chapters 1 and 2. By the time of the outbreak of World War I he was one of the foremost observational astronomers of the day, and had recently been elected a Fellow of the Royal Society and appointed the Plumian Professor at Cambridge University. The war had effectively cut off direct communication between British and German scientists, but the Dutch scientist Willem de Sitter managed to forward to Eddington Einstein's latest paper together with several of his own on the general theory of relativity. Eddington recognized the deep implications of this new theory, and he immediately set out to learn the mathematics required to master it. In 1917 he prepared a detailed report on the general theory for the Physical Society of London. This helped spread the word.

Eddington and Astronomer Royal Frank Dyson also began to contemplate an eclipse expedition to measure the predicted deflection of light. As an astronomer at the Royal Greenwich Observatory from 1906 to 1913, Eddington had made an eclipse expedition in 1912 to study features of the Sun, such as the solar corona, and was familiar with the techniques and problems involved. Dyson had pointed out that the eclipse of 29 May 1919 would be an excellent opportunity because of the large number of bright stars expected to form the field around the Sun. A grant of 1,000 pounds sterling (around 70,000 US dollars today) was obtained from the government, and planning began in earnest. The outcome of the war was still in doubt at this time, and a danger arose that Eddington would be drafted. As a devout Quaker and ardent pacifist he had pleaded exemption from military service as a conscientious objector, but, in its desperate need for more manpower, the Ministry of National Service appealed the exemption. Finally, after three hearings and a last-minute appeal from Dyson attesting to Eddington's importance to the eclipse expedition, the exemption from service was upheld on 11 July 1918. This was just one week before the second Battle of the Marne, a pivotal event in that war. Eddington also firmly believed, perhaps naively, that the example of British scientists verifying the theory of a German physicist would demonstrate how science could lead the world toward peace.

On 8 March 1919, just four months after the end of hostilities, two expeditions set sail from England. After a brief stop at the island of Madeira, the teams split up. Eddington, accompanied by Edwin Cottingham, headed for the island of Principe, off the coast of present-day Equatorial Guinea; Charles Davidson and Andrew Crommelin headed for the city of Sobral, in northern Brazil. The principle of the experiment is deceptively simple. During a total solar eclipse, the Moon hides the Sun completely, revealing the field of stars around it. Using a telescope and photographic plates, the astronomers take pictures of the obscured Sun and the surrounding star field. These pictures are then compared with pictures of the same star field taken when the Sun is not present. The comparison pictures are taken at night, weeks or months before or after the eclipse, when the Sun is nowhere near that part of the sky and the stars are in their true, undeflected positions. In the eclipse pictures, the stars whose light is deflected would appear to be displaced away from the Sun relative to their actual positions (see Figure 3.2).

One property of the predicted deflection is important: Although a star whose image is at the edge of the Sun is deflected by 1.75 arcseconds, a star whose image is twice as far from the center of the Sun is deflected by half as much, and a star ten times as far is deflected by one tenth; in other words, the deflection varies inversely as the angular distance of the star from the Sun (see Figure 3.2). Now, because the eclipse pictures and the comparison pictures are taken at different times, under different conditions (and sometimes using different telescopes), their overall magnifications may not be the same. Therefore, the stars in the photographs that are farthest from the Sun, undeflected on the comparison plate, deflected only negligibly on the eclipse plate, can be used to determine an overall magnification correction. Then, the true deflection of the stars closest to the Sun can be measured.

In practice, of course, nothing is ever this simple. One important complication is a phenomenon that astronomers call "seeing." Because of turbulence in the Earth's atmosphere, starlight passing through it can be refracted or bent by the warmer and colder pockets of moving air and can suffer deflections of as much as a few arcseconds (this is part of what makes stars twinkle to the naked eye). These deflections are comparable to the effect being measured. But because they are random in nature (as likely to be toward the Sun as away from it), they can be averaged away if one has many images. The larger the number of star images, the more accurately this effect can be removed. Therefore, it is absolutely crucial to obtain as many photographs with as many star images as possible. To this end, of course, it helps to have a clear sky.

We can therefore imagine Eddington's emotional state when, on the day of the eclipse, "a tremendous rainstorm came on." As the morning wore on, he began to lose all hope. Before the expedition, Dyson had joked about the possible outcomes: no deflection would show that light was not affected by gravity, a half deflection would confirm Newton, and a full deflection would confirm Einstein. Eddington's companion on Principe had asked Dyson before the departure what would happen if they found double the deflection. Dyson had answered, "Then, my dear Cottingham, Eddington will go mad, and you will have to come home alone." Now Eddington had to consider the possibility of getting no results at all. But at the last moment, the weather began to change for the better: "The rain stopped about noon, and about 1:30, when the partial phase [of the eclipse] was well advanced, we began to get a glimpse of the Sun." Of the sixteen photographs taken through the remaining cloud cover, only two had reliable images, totaling only about five stars. Nevertheless, comparison of the two eclipse plates with a comparison plate taken at the Oxford University telescope before the expedition yielded results in agreement with general relativity, corresponding to a deflection for a grazing ray of 1.60 ± 0.31 arcseconds, or 0.91 ± 0.18 times the Einsteinian prediction. The Sobral expedition, blessed with better weather, managed to obtain eight usable plates showing at least seven stars each. The nineteen plates taken on a second telescope turned out to be worthless because the telescope apparently changed its focal length just before totality of the eclipse, possibly as a result of heating by the Sun. Analysis of the good plates yielded a grazing deflection of 1.98 ± 0.12 arcseconds, or 1.13 ± 0.07 times the Einsteinian value.

Eddington made the announcement of the measurements at a joint meeting of the Royal Society of London and the Royal Astronomical Society on 6 November 1919. He may be the first scientist to fully appreciate the power of the media of his day, and engineered some adroit advance publicity. The mathematician Alfred North Whitehead described the scene: "The whole atmosphere... was exactly that of a Greek drama... in the background the picture of Newton to remind us that the greatest of scientific generalizations was now, after more than two centuries, to receive its first modification." Before this, Einstein had been an obscure Swiss/German scientist, well known and respected within the small European community of physicists, but largely unknown to the outside world. With newspaper headlines spreading worldwide during the following days, everything changed, and Einstein and his theory became immediate sensations. The Einstein aura has not abated since.

On the other hand, Einstein's fame did engender a backlash, especially in Germany. The rise of nationalism and anti-Semitism in Germany between the world wars had its counterpart in scientific circles. In 1920, Paul Weyland organized a public forum in which Einstein and his theories were denounced. One of the leading exponents of this view was Philipp Lenard, a Nobel Laureate in Physics (1905) for his work on cathode rays (electron beams in modern parlance). An avowed sympathizer of the nascent Nazi movement, Lenard spent much of his time between the wars attempting to cleanse German science of the "Jewish taint." Relativity represented the epitome of "Jewish science," and much effort was expended by Lenard and others in attempts to discredit it. In early 1921, while preparing an article against general relativity,

Lenard learned of the existence of Georg von Soldner's 1804 paper. This discovery delighted him, because it showed the precedence of von Soldner's "Aryan" work over Einstein's "Jewish" theory. The fact that the eclipse results appeared to favor Einstein over von Soldner did not appear to faze him. Lenard prepared a lengthy introductory essay, incorporated the first two pages of von Soldner's paper verbatim and summarized the rest, and had the whole thing published under von Soldner's name in the 27 September 1921 issue of the journal *Annalen der Physik*.

The vast majority of non-Jewish German physicists did not share these views, however, and despite the Nazi takeover in Germany and the subsequent dismissal and emigration of many Jewish physicists (including Einstein), the anti-relativity program became little more than a footnote in the history of science.

There were legitimate scientific questions about Eddington's results, however. Given the poor quality of the data, did they really support Einstein or not? In 1980, some historians of science wondered whether Eddington's enthusiasm for the theory of general relativity caused him to select or massage the data to get the desired result. Numerous reanalyses between 1923 and 1956 of the plates used by Eddington yielded the same results as he obtained within 10 percent. In 1979, on the occasion of the centenary of Einstein's birth, astronomers at the Royal Greenwich Observatory near London reanalysed both sets of Sobral plates using a modern tool called the Zeiss Ascorecord and its data reduction software. The plates from the first Sobral telescope yielded virtually the same deflection as that obtained by Davidson and Crommelin, with the errors actually reduced by 40 percent. Despite the scale changes in the plates from the second Sobral telescope, the analysis still gave a result 1.55 ± 0.34 arcseconds for a grazing ray, consistent with general relativity, albeit with much larger errors, reflecting the problem with the telescope focal length. Looking back on the British astronomers' treatment of the data, our colleague Daniel Kennefick has argued that there is no credible evidence of bias on their part.

But scientists are reluctant to adopt a world-changing theory on the basis of the measurements of a single team. Any new theory of nature must stand the test of many experimental checks by different groups with different methods and techniques. Strangely, one set of measurements made *before* the 1919 eclipse failed to confirm Einstein's prediction. William Campbell and Heber Curtis of the Lick Observatory analyzed plates from a 1900 eclipse near Augusta, Georgia and a 1918 eclipse at Goldendale, Washington in the USA, hoping to beat the British to the punch. Unfortunately the quality of the images was poor, and they found no unambiguous evidence for the Einstein deflection; ironically, Campbell reported this negative result at the Royal Astronomical Society meeting on 11 July 1919 while Eddington was still at sea returning from Principe. At the meeting, Dyson reported that Eddington had telegraphed that his prelimary measurements indicated a positive result.

Following up on Eddington's success, seven teams tried the measurement during a 1922 eclipse in Australia, although only three succeeded in getting usable data. Campbell and Robert Trumpler of the Lick team reported a result for the grazing deflection of 1.72 ± 0.11 arcseconds, while a Canadian team and an English/Australian team reported values between 1.2 and 2.3 arcseconds. Later eclipse measurements continued to support general relativity: one in 1929, two in 1936, one in 1947, one in 1952 and one in 1973. Surprisingly, there was very little improvement in accuracy, with different measurements giving values ranging as far as 30 percent away from the Einstein value. Still, there was little doubt that Einstein beat Newton.

The 1973 expedition is a case in point. Organized by the University of Texas and Princeton University, the observation took place in June at Chinguetti Oasis in Mauritania. The observers had the benefit of 1970s technology: Kodak photographic emulsions, a temperature-controlled building housing the telescope (the outside temperature at mid-eclipse was 97°F), sophisticated motor drives to control the direction of the telescope accurately, and computerized analysis of the photographs. Unfortunately they couldn't control the weather any better than Eddington could. Eclipse morning brought high winds, drifting sand, and dust too thick to see the Sun. But as totality of the eclipse approached, the winds died down, the dust began to settle, and the astronomers took a sequence of photographs during what they have described as the shortest six minutes of their lives. They had hoped to gather over 1,000 star images, but the dust cut the visibility to less than 20 percent and only a disappointing 150 were obtained. After a follow-up expedition to the site in November to take comparison plates, the photographs were analyzed using a special automated device called the GALAXY Measuring Engine at the Royal Greenwich Observatory. The result agreed with the Einsteinian prediction within the measurement error of about 10 percent, still only a modest improvement over previous eclipse measurements.

This was the backdrop for Don Bruns' attempt to do an improved eclipse measurement in 2017. Bruns had retired in 2014 after a career in the optics industry, working on lasers and advanced optics for both military and commercial applications. He knew astronomical instrumentation inside and out, and decided to employ twenty-first-century technology in his attempt to redo this historic measurement. Among his advantages were the CCD camera, promising greatly improved response to the incoming starlight and improved image stability over photographic emulsions or the glass plates used by Eddington. He also did not have to worry about taking comparison images of the star field before or after the experiment, because an orbiting telescope known as Gaia, launched in 2013, was providing undeflected positions of all the relevant stars with an accuracy far better than he could ever obtain himself. Finally, the telescope and camera could be completely controlled by a computer using software written, tested and rehearsed in advance. In fact, unlike the many teams before him, Bruns reported that he could actually sit back and enjoy the eclipse, because everything was pre-programmed. Excellent weather didn't hurt. Nevertheless, he had to sweat many tedious details in his analysis of the data before he could report a value of 1.75 arcseconds for a grazing ray, with a probable uncertainty of 3 percent, in excellent agreement with general relativity and with about three times smaller uncertainty than Eddington had claimed.

Bruns' measurement is of mainly historical and personal interest, because by the late 1960s testing of Einstein's deflection during solar eclipses was already being superseded by a technique that was a marriage of two of the most important astronomical discoveries of the twentieth century: the radio telescope and the quasar.

Radio astronomy began in 1931, when Karl Jansky of the Bell Telephone Laboratories in New Jersey found that the noise in the radio antenna he was trying to improve for use in radio telecommunications was coming from the direction of the center of our galaxy (we will return to this in Chapter 6). The development of radar during World War II led to new receivers and techniques, and to the rapid development of radio telescopes as new astronomical tools. Among the sources of radio waves that were discovered were the Sun itself, interstellar gas clouds such as the Crab Nebula, clouds of hydrogen atoms and of complex molecules, and radio galaxies. Radio waves are the same as ordinary visible light, only of longer wavelength. Whereas visible light spans a wavelength range from 400 to 700 nanometers (a nanometer is a billionth of a meter), radio waves span the range from a tenth of a millimeter to several meters. General relativity predicts exactly the same deflection of radio waves as visible light; the effect is independent of wavelength.

To measure the deflection of radio waves, we need to be able to measure to high precision the direction from which they come. To this end, the radio interferometer is the ideal instrument. In its simplest form, a radio interferometer consists of two radio telescopes separated by some distance, called the baseline (Figure 3.5). As a radio wave from some external source approaches the pair, the wavefronts may arrive at one telescope before they arrive at the other, depending on the location

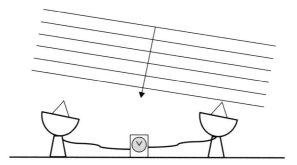

Fig 3.5 Radio interferometry. A radio wave approaches two radio telescopes. Each wave arrives first at one telescope, then at the other. The times are compared very precisely using an atomic clock linked to the two telescopes, leading to accurate determinations of the direction of the source.

of the source on the sky. The difference in the time of arrival of a given wave front at the two telescopes is measured by comparing the signals using a precise atomic clock. For a given wavelength of the radio waves, the longer the baseline, the larger the time delay for a given angle of approach, and thus the more accurately the angle can be measured. Radio interferometers range in baseline from the 1 kilometer instrument in Owens Valley, California to the 42 kilometer long "Y" containing twenty-seven linked antennae along its three legs at the Very Large Array in New Mexico, to the Event Horizon Telescope (EHT), which links antennae as far apart as Hawaii, Chile, Europe and the South Pole in a global interferometer (we will return to EHT in Chapter 6). When the telescopes are separated by transcontinental and intercontinental distances, the technique is known as Very Long Baseline Interferometry (VLBI). The resolution of some of these VLBI interferometers can be better than one ten-thousandth of an arcsecond, or 100 microarcseconds. That would be good enough to resolve this book from Earth if it were sitting on the surface of the Moon.

We also need a very sharp source of radio waves. Most astronomical sources are unsuitable for this purpose because they are extended in space. For example, most galaxies that emit radio waves (or radio galaxies, for short) do so from an extended region that can be as large as a degree in angular size. The discovery of quasistellar radio sources, or quasars as they are called, besides motivating applications of general relativity to astrophysics, provided the ideal source of radio waves to test the deflection of light. Because they are so distant, between one and twelve billion light years away, they appear much smaller in extent, making it possible to pinpoint their locations more accurately. Yet despite their distance, many of them are powerful radio sources and their light emission is constant enough to enable long-term observations.

Unfortunately, a powerful point source of radio waves is not the only ingredient for a successful light deflection experiment. We need at least two of them fairly close to each other on the sky, and they have to pass near the Sun as seen from Earth. We need at least two for the same reason as we needed a field of stars behind the eclipsed Sun in the optical deflection measurements: the stars whose images are far from the Sun are used to establish the scale because their light is relatively undeflected, and the movement of the star images close to the Sun is used to determine the deflection. Figure 3.6 illustrates how this would work. The Sun passes in front of a pair of quasars, one about 1 degree away, the other about 4 degrees away (top panel). Initially the angle between the two quasars as measured on Earth is the nominal, unperturbed angle (bottom panel). As the Sun approaches the lower quasar, the quasar's image as seen from Earth is displaced toward the other quasar, causing the angle between them to decrease. Then, as the Sun passes the lower quasar, its image is displaced to the left, away from the other quasar, causing the angle between them to increase, although less dramatically. As the Sun moves away from the pair, the angle returns to its nominal value.

Early measurements took advantage of the fact that groups of strong quasars annually pass very close to the Sun (as seen from the Earth), such as the group 3C273, 3C279 and 3C48 (the designation "3C" refers to the Third Cambridge Catalogue of radio sources). As the Earth moves in its orbit, changing the lines of sight of the quasars relative to the Sun, the angular separation between pairs of quasars varies. A number of measurements using radio interferometers over the period 1969–1975 yielded accurate determinations of the deflection, reaching levels of 1 percent.

In recent years, scientists interested in the Earth have made their own use of VLBI. The idea is to measure directions to hundreds of radio galaxies and quasars everywhere in the sky in order to monitor very precisely the Earth's rotation rate and the orientation of its rotation axis. Variations in the rotation of the Earth can be caused by changes in ocean levels, variations in weather patterns, interactions between the Earth's mantle and its core, and the gravitational tug of the Moon. A test of relativity is but a by-product of their work. But because the intrinsic accuracy of the measurements is now so high, they are sensitive to the deflection of light over almost the entire celestial sphere. While a ray grazing the Sun is deflected by 1.75 arcseconds, a ray approaching the Earth from 90° relative to the Sun is deflected by 0.004 arcseconds or 4 milliarcseconds. Even a ray coming from 175°, almost directly opposite the direction of the Sun, is deflected by just under a milliarcsecond.

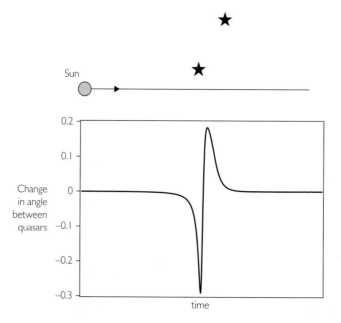

Fig 3.6 Testing the deflection of light using quasars. Top panel: As seen from Earth, the Sun moves across the sky passing near the sky position of two quasars. When the Sun is far to the left, the angle between the two quasars is the nominal, unperturbed angle. As the Sun approaches the position of the lower quasar, the quasar's image is displaced toward the other quasar, causing the angle between them to decrease. Then, as the Sun continues past the lower quasar, its image is displaced to the left, away from the other quasar, causing the angle between them to increase. As the Sun moves far to the right, away from the pair, the angle returns to its nominal value. Bottom panel: The changes in angle between the quasars plotted against time.

With accuracies of 10 to 100 *micro*arcseconds, modern VLBI can detect these tiny deflections. Recent global analyses of several million VLBI observations of over 500 quasars and compact radio sources, made by telescopes spread around the globe, confirmed general relativity to about 0.01 percent, or one part in 10,000. The vast majority of the sources were more than 30 degrees from the Sun at all times. It is no longer necessary to look right at the Sun to detect Einstein's light deflection effect!

At the time of the writing of this book, radio astronomers seem to have the upper hand in testing the deflection of light, but not for long. Optical astronomers may yet have the last laugh. The idea is to put

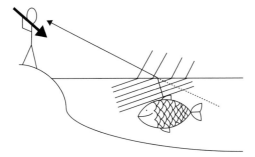

Fig 3.7 Refraction of light. Because of the change of the speed of light in going from water to air, the spear fisher must compensate for the bending or refraction of the light rays from the fish (obviously not drawn to scale!).

Eddington into space, hypothetically speaking of course, to get above the effects of the Earth's atmosphere. This was first demonstrated by the Hipparcos satellite, launched in 1989 by the European Space Agency (ESA) and operated until 1993. Hipparcos made precise measurements of the positions of over two million stars at optical wavelengths and was able to test the distorting effect of the deflection of light on the celestial sphere to about one part in 1,000, in agreement with general relativity, but not quite as precise as VLBI. But its follow-up mission, called Gaia, launched by ESA in 2013, is making even more precise position measurements of about a *billion* stars. This may permit a test of general relativity's light bending to one part in a million.

Astronomers of the ancient world referred to the stars as residing on a "celestial sphere" that was fixed and immutable. This stellar realm *had* to be so perfect, because it was where the gods resided. We now know, thanks to Einstein, that it is more like a soap bubble; as the Sun wanders across the sky, the celestial sphere appears to warp and distend as light from those distant stars wends its way through the piece of curved spacetime that the Sun carries with it. Even after dark, with the Sun behind us, the night sky is warped. At fractions of a milliarcsecond, the effect is far too small for our eyes to sense, but astronomers now measure it routinely.

But this is not the only effect of gravity on light. Gravity also slows light down. As anyone knows who has ever tried and failed to spear a fish swimming in a river or a lake, there is a close relationship between the bending of light and changes in the speed of light. The speed of light in water is about 75 percent of its speed in air, because the light's progress is impeded by its interactions with the atoms of the denser water, just as a person takes longer to get across a crowded room than an empty room. This means that as the waves of light reflect off the surface of the fish and cross the interface with the air, they begin to travel faster. This causes the wave fronts to be tilted more toward the vertical (Figure 3.7). To the observer on the shore, the apparent direction of the fish is defined by the verticals to the wavefronts, and thus the fish appears to be above its true location. This is why your spear misses the fish, unless you compensate for this effect, known as refraction.

Therefore, if the curved spacetime around the Sun causes light to bend, then there must be an associated change in its speed.

But wait a minute, this can't be right! According to the equivalence principle, the speed of light as measured in any local freely falling frame is always the same. How then can we say that the light slows down near the Sun?

The problem here is the distinction between local effects, those that are observable in one very small, freely falling frame, and large-scale or global effects, which cover a range of space or an interval in time large enough that the effects of curvature of spacetime are important and cannot be described by a single freely falling frame. One indication of the global nature of an effect like the deflection of light was the fact that we could not detect it by looking at a single star or quasar; we always had to compare the light from one star or quasar with that from another that appeared to be farther from the Sun.

Similarly with the speed. An observer in a small, freely falling space-ship close to the Sun will find that the speed of light, given by the width of her ship divided by the time taken for the ray to cross it, is exactly the same as the speed obtained by a similarly freely falling observer far from the Sun. But the rates of the clocks of the two observers are not the same, because of the gravitational redshift discussed in Chapter 2, and the rulers they use to measure distances are not the same because of the warpage of space, as represented in our two-dimensional sheet of

Figure 3.4. Thus, if we were to add up all the times taken for the ray to pass through a sequence of such frames laid side to side, we would find that the total travel time for the ray and frames that are close to the Sun is slightly longer than the time for a similar ray and frames that pass nowhere close to the Sun. Our rubber sheet of Figure 3.4 also suggests a delay, since the ray must "dip" down as it follows the rubber surface, thus taking longer to get across compared to the ray that passes far from the ball (remember that this is just an imperfect analogy!).

To make this slightly more concrete, consider Figure 3.8. An enormous circular rigid ring has been constructed with a diameter much larger than the solar system, with an emitter of light on one side of the ring and a receiver on the opposite side. The ring is so large that the Sun's gravity has no measurable effect on it. The Sun is moving relative to the ring in such a way that it will pass through the center of the ring (clearly this

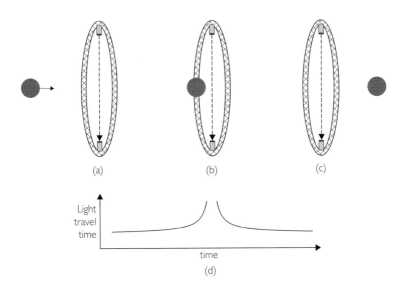

Fig 3.8 Excess travel time of light. An emitter on an enormous rigid ring sends light signals to a receiver on the opposite side. (a) The Sun is far to the left, so the travel time is the diameter of the ring divided by the usual speed of light. (b) The Sun is about to pass through the center of the ring, so the signal suffers an excess delay caused by curved spacetime near the Sun. (c) The Sun is far to the right, so the travel time returns to normal. (d) Plot of the travel time vs. time.

is a *gedanken* experiment!). In (a), the Sun is initially on the left, very far from the center of the ring. The Sun's gravity has a negligible effect on the propagation of the light ray, so that the time the ray takes to cross the ring is simply the diameter divided by the speed of light. In (b), the Sun is close to the center of the ring, and the light ray passes close to the Sun. According to our argument above, the ray takes a little longer to cross the ring.[1] In (c), the Sun has passed to the other side, and the time to cross the ring returns to its normal value. Plot (d) shows schematically the increase in travel time as a function of time (the gap in the middle is where the light rays hit the Sun and never reach the other side).

Whether or not we use the words "light slows down near the Sun" is purely a question of semantics. Because the observer on the ring who receives the light ray never goes near the Sun to make the measurement, she can't really make such a judgment; and if she had made such a measurement in a freely falling laboratory near the Sun, she would have found the same value for the speed of light as in a freely falling laboratory far from the Sun, and might have thoroughly confused herself. All the observer can say with no fear of contradiction is that she observed an excess time of travel that depended on how close the light ray came to the Sun. The only sense in which it can be said that the light slowed down is mathematical: in a particular mathematical representation of the equations that describe the motion of the light ray, what general relativists call a particular coordinate system, the light appears to have a variable speed. But in a different mathematical representation (a different coordinate system), this statement might be false. Nevertheless, the observable quantities, such as the net time of travel, are the same no matter what representation is used. This is one of those cases in relativity where the careless use of words or phrases that are not based on observable quantities can lead to confusion or contradiction. We have already seen an example of this in our discussion in Chapter 2 of what

---

[1] The attentive reader might ask if the deflection of the light ray adds to the distance traveled, and hence to the time. Indeed it does, but that effect is negligible compared to the effect we are describing.

"really" changed in the gravitational redshift, the clock rates or the signal frequency.

Given this discussion, you might be tempted to assume that Einstein derived this delay effect and proposed that it be measured. But he didn't. The effect was derived by the radio astronomer Irwin I. Shapiro in 1964.

After receiving a Ph.D. in physics from Harvard University in 1955, Shapiro had been at MIT's Lincoln Laboratory working on the problem of "radar ranging." The idea was to bounce radar signals off planets such as Venus and Mercury and measure the round-trip travel times. This new technique had already led to improved determinations of the astronomical unit, the mean radius of the Earth's orbit, and promised significant improvements in determining the orbits of the planets.

Shapiro had only a passing acquaintance with general relativity, and might not ever have considered it relevant to radar ranging had it not been for a lecture he attended in 1961 on measurements of the speed of light. Purely in passing, the speaker mentioned that according to general relativity, the speed of light is not constant. This statement puzzled Shapiro, because he had always thought that according to relativity, the speed of light should be constant. He knew, of course, that general relativity predicts that light should be deflected by a gravitating body, and following the same logic as we presented for the fish in water, he asked if its speed would also be affected.

To be fair, Einstein had considered the possibility of speed variation. Once he understood, from the principle of equivalence, that gravity could have an effect on light (the gravitational redshift), he attempted to construct a theory of gravity in which the speed of light would vary in the vicinity of a gravitating body. It was the equations from this specific theory that he used in 1911 to calculate the (one-half) bending of light. As we discussed on page 42, we're talking here about the speed in a specific coordinate system.

However, for some reason Einstein did not take the next step, the one that Shapiro took. Shapiro consulted the classic general relativity textbook by Eddington and found that, according to the equations of the full general theory, the "effective" speed of light should indeed vary, just as it did in Einstein's earlier model (in the full theory of general relativity the effect was doubled, just as it was for the deflection of light). Shapiro then applied these equations to the problem of the round trip of a radar signal to a distant object and found, in agreement with our qualitative argument, that the radar signal should take slightly longer to make the round trip than one would have expected on the basis of Newtonian theory and a constant speed of light. The additional delay would increase if the signal passed closer to the Sun.

In the solar system, the effect would be most noticeable when the target was on the opposite side of the Sun from the Earth, so that the signal would pass very near the Sun on its trip, as in panel (b) of Figure 3.8. Such a configuration is called superior conjunction (when both planets are on the same side of the Sun, it is called inferior conjunction). For example, Shapiro found that a radar signal sent from Earth to Mars at superior conjunction that just grazes the surface of the Sun suffers a round-trip delay of 250 millionths of a second (250 microseconds). Don't forget that the total round-trip travel time for such a signal is about 42 minutes! So the idea here would be to detect an additional delay in the round trip of 250 microseconds on a total travel time of three-quarters of an hour. This might seem to be a hopeless proposition until you realize that the distance that light travels in 250 microseconds is 75 kilometers. So the delay represents an apparent shift in the distance to the target of half of this, or about 38 kilometers. Since Shapiro saw that radar ranging could potentially achieve a precision in distance between Earth and planets corresponding to a few kilometers, then perhaps this effect could be observed. The problem was that no radio telescopes at the time had the capability of sending a powerful enough radar signal to any planet at superior conjunction and detecting the extremely weak return signal. So Shapiro's calculation lay in his desk for two years.

In the fall of 1964, two events caused Shapiro to retrieve his superior conjunction calculation and take it more seriously. The first was the completion of the Haystack radar antenna in Westford, Massachusetts. The second was the birth of his son on 30 October. As often happens in creative endeavors, the event in his personal life may have elevated him to a higher level of awareness or of mental activity, for soon thereafter, while describing the time delay idea to a colleague at a party, he suddenly realized that Haystack might be able to range to Mercury at superior conjunction and provide a means to test the time delay prediction (Mars would be too far away at superior conjunction for Haystack to record a measurable signal). Shapiro decided then to write up his superior conjunction calculation for *Physical Review Letters*. The paper was submitted in the middle of November, and published under the title "Fourth Test of General Relativity" in late December, 1964. (The first three tests were the gravitational redshift, the light deflection and the perihelion advance of Mercury, the three proposed by Einstein.) In time, the effect would come to be called the Shapiro time delay.

The principle behind the measurement of the time delay is very much the same as the principle behind the measurement of the deflection of light. Just as we could not measure the deflection of a single star, we cannot detect the time delay in a single radar shot. The reason, of course, is that we cannot "turn off" the gravitational field of the Sun in order to see what the star's "true" position is or to see what the "flat spacetime" round-trip travel time would have been. To get at the deflection, we had to compare the position of a star or quasar relative to other stars or quasars both when its light passed far from the Sun, and when its light passed very near the Sun. By the same token, to see the time delay, we must compare the round-trip travel time of a radar signal to the planet when the signal passes far from the Sun with that when the signal passes close to the Sun.

When the signal to the planet passes far from the Sun, the Shapiro time delay is relatively small, and the round-trip travel time is closer to being a measure of the "true" distance. This corresponds to the situation in Figure 3.4 where the signal traverses a portion of space that is virtually flat. As the planet moves into superior conjunction, however, and the signal passes closer and closer to the Sun, the Shapiro time delay becomes a larger contribution to the round-trip travel time.

However, even though the radar signal may go near the Sun, the planet itself never does. Its orbit is well away from the Sun, on the order of 230 million kilometers for Mars or 58 million kilometers for Mercury, for instance. Because of this, the planet always moves through a region of low spacetime warpage, and maintains a relatively low velocity; therefore, the relativistic effects on its orbit are small. To the accuracy desired for a time delay measurement, its orbit can be described quite adequately by standard Newtonian gravitational theory. Therefore, even though the planet moves during the experiment, its motion can be predicted accurately. Because of this circumstance, the time delay can be measured in four steps: (1) by ranging to the planet for a period of time when the signal stays far from the Sun, determine the parameters that describe its orbit at that time; (2) using the orbit equations of Newtonian theory, including the perturbations from all the other planets, make a prediction of its future orbit and that of the Earth, including especially the period of superior conjunction where the action will occur; (3) using the predicted orbit, calculate the round-trip travel times of signals to the planet assuming no Shapiro time delay; and (4) compare these predicted round-trip travel times with those actually observed during superior conjunction, attribute the difference to the Shapiro time delay, and see how well it agrees with the prediction of general relativity.

Within about a month of submitting his paper on the time-delay effect to *Physical Review Letters*, Shapiro's colleagues at Lincoln Laboratory set out to upgrade the Laboratory's Haystack radar by increasing its power fivefold and by making other electronic improvements. This would give them the capability to get a decent echo from Mercury and also Venus at superior conjunction, and to measure the round-trip travel times to within 10 microseconds. By late 1966 the improved system was ready, just in time for the 9 November superior conjunction of Venus. Unfortunately, Venus goes through superior conjunction only about once every year and a half, so after observing Venus they then turned the radar sights on Mercury. Because Mercury orbits the Sun almost three times faster than Venus, it has a superior conjunction more often, about three times per year, giving more opportunities to measure the time delay. Measurements were made during the 18 January, 11 May and 24 August 1967 conjunctions of Mercury. All told, over four hundred radar "observations" were used. Most of these measurements (the ones not taken near superior conjunction of either of the planets) were combined with existing optical observations of Mercury and Venus available through the US Naval Observatory to accomplish the first step of the method, namely, to establish accurate orbits for the two planets. The remaining radar measurements centered around the superior conjunctions were then used to compare the predicted time delays with the observed time delays (because of large amounts of noise, the Venus data turned out not to be very useful). The results using Mercury data agreed with general relativity to within 20 percent. The first new test of Einstein's theory since 1915 was a reality.

But the story does not end there. During the summer of 1965, while Shapiro and his colleagues were busy working on the Haystack radar, a US spacecraft hurtled past Mars, the first man-made object to encounter the "red planet." The spacecraft was Mariner 4, and on its way by the planet it took twenty-one pictures and examined the Martian atmosphere using radio waves. Buoyed by the success of Mariner 4, NASA in December 1965 authorized two more missions to Mars, Mariner 6 and 7 in 1969 (Mariner 5 was a Venus mission) and Mariner 8 and 9 in 1971, and planners began to think seriously about Martian landers. While these missions would bring planetary exploration to a zenith, at least temporarily, they would also have crucial consequences for general relativity.

At the Jet Propulsion Laboratory (JPL) in Pasadena, California, where the Mariner program was headquartered, the relativistic time delay was also on people's minds, and they began to wonder if there was any way to make use of Mariner 6 and 7 to measure the time delay. In fact, two JPL scientists, Duane Muhleman and Paul Reichley, had calculated the delay effect of general relativity on radar propagation independently of Shapiro, although they only published the results in internal JPL reports. There was no reason in principle why a measurement of the delay should not be possible. Other than in size, there is no fundamental difference between a planet and a spacecraft. The orbit of the spacecraft can be determined by tracking, and its trajectory during superior conjunction can be predicted, just as for ranging to the planet, and the time delay of the radar ranging signal during superior conjunction can be measured and compared with the prediction of general relativity.

Mariners 6 and 7 were launched on 24 February and 27 March 1969, and reached Mars by the end of July. Both spacecraft performed their primary tasks of observing Mars' surface and atmosphere beautifully, and then left the planet to go into orbit around the Sun. Between December 1969 and the end of 1970, several hundred range measurements were made to each spacecraft, with the heaviest concentration, involving almost daily measurements, around the time of each superior conjunction—on 29 April 1970 for Mariner 6 and on 10 May for Mariner 7. Neither spacecraft actually went behind the Sun. Because of the tilt of their post-Martian orbits, they both passed by the Sun slightly to the north, Mariner 6 about 1° away, Mariner 7 about 1.5° away, as seen from Earth. For Mariner 6, the distance of closest approach of the radar signal at superior conjunction was about 3.5 solar radii, corresponding to a Shapiro time delay of 200 microseconds out of a total round-trip travel time of 45 minutes. For Mariner 7, the radar signals came no closer than about 5.9 solar radii, giving a slightly smaller time delay of 180 microseconds. After feeding all the observations into the computer, they found that the measured delays agreed with the predictions of general relativity to within 3 percent. This was a dramatic improvement over the 20 percent figure from Venus and Mercury ranging.

Of course, the planetary radar ranging people at Lincoln Laboratory had not been idle since 1967. They had continued to make radar observations of Mercury and Venus using both the Haystack antenna and the Arecibo radio telescope in Puerto Rico. In fact, during late January and early February 1970, while the JPL rangers were busy getting distances to the Mariner spacecraft on their approach to superior conjunction, Venus passed through its own superior conjunction, bombarded almost twice a week by radar signals from Haystack and Arecibo. Data from that Venus conjunction, and from the numerous Mercury conjunctions between 1967 and the end of 1970, once again yielded relativistic time delays in agreement with general relativity, this time at the 5 percent level.

It was soon realized, however, that each method, planetary vs. spacecraft tracking, had advantages and disadvantages. One advantage of planets is that they are massive and therefore are completely unaffected by the constant bombardment of the solar wind and solar radiation pressure. Spacecraft, by contrast, are light and have large antennae and solar panels, and so they tend to get jostled around a lot on their way through the rough neighborhood of interplanetary space. This is important because of the need to predict the orbit accurately during the time of superior conjunction when range measurements are supposed to yield the Shapiro delay.

An advantage of spacecraft is that they receive the radar signal from Earth, pass it through a transponder (the same device we encountered in Chapter 2), which boosts the signal's power and beams it right back to Earth, leading to very accurate round-trip travel times. By contrast, planets are poor reflectors of radar beams, and also have valleys and mountains that introduce uncertainties in the "true" round-trip travel time.

The way to combine the transponding capabilities of spacecraft with the imperturbable motions of planets was to anchor a spacecraft to a planet, by having the spacecraft orbit the planet, or even better by letting the spacecraft land on the planet.

The first anchored spacecraft was Mariner 9, the orbiter that reached Mars in November 1971 just in time to photograph the raging dust storm that obliterated most of the planetary surface for several weeks. The next Martian superior conjunction of 8 September 1972 gave a confirmation of the Einsteinian time delay to 2 percent, only a modest improvement over the previous results, but enough to prove the power of the anchoring idea.

And then came Viking. The Viking landers on Mars were a spectacular achievement for planetary exploration, with their close-up views of the Martian surface, their analyses of the atmosphere, and their search for signs of life in the Martian soil. But to the general relativist they were even more beautiful, for they were the perfect anchored spacecraft for the time delay experiment. After a ten-month voyage, the first Viking spacecraft reached Mars in mid June, 1976. After several weeks studying possible landing sites, Lander 1 was detached from the orbiter and descended to a plain called Chryse on 20 July. Eighteen days later, the second Viking reached Mars, and on 3 September, Lander 2 dropped to the surface in a region called Utopia Planitia.

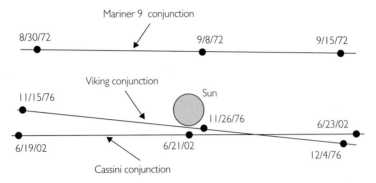

Fig 3.9 Paths of Mariner 9, Viking and Cassini during superior conjunctions, as seen from Earth. Although Cassini was almost 8.5 astronomical units away on the far side of the Sun, the tracking signal passed within 0.6 solar radii of the Sun's surface. Even though the signal from the Viking landers passed closer to the Sun at superior conjunction, by the time of Cassini, improvements in radar tracking and atomic clocks, plus Cassini's quiet orbit en route to Saturn, led to a much better test of the Shapiro delay.

While much of the world focused its attention on the remarkable photographs and scientific data radioed back to Earth by the landers and orbiters, Shapiro's group at MIT and the JPL team, now working together, began to prepare for the 26 November superior conjunction. With two landers and two orbiters all providing ranging data, they had an excellent configuration of anchored spacecraft so as to avoid the errors of random orbit perturbations.

Ranging measurements were made from the moment the landers reached the Martian surface, through the November superior conjunction, and on until September 1977, when the two groups felt they had enough data to measure the Shapiro delay. The final result was a measured time delay in complete agreement with the prediction of general relativity, with an accuracy of 0.1 percent, or one part in a thousand!

Another twenty-five years would pass before the next leap forward in testing the Shapiro delay: the Cassini-Huygens mission. And what a leap it was. Launched jointly by NASA and ESA on 15 October 1997 and terminating on 15 September 2017, the mission is best known to the general public for its extraordinary feats: detection of new features in the atmosphere of Jupiter during its flyby of that planet en route to Saturn;

discovery of seven new moons of Saturn; close flybys of many Saturnian moons, including Titan, Phoebe and Enceladus; the first spacecraft to orbit Saturn; the successful landing of the Huygens probe on Titan; and the grand finale suicide dive of Cassini into Saturn's atmosphere, sending back useful data until the very end.

What is less well known is that Cassini made possible a test of the Shapiro delay and general relativity to a precision of one part in 100,000, a hundred times better than Viking. Several things made such an accurate test possible. On 21 June 2002 Cassini was in "cruise" mode en route to Saturn, about 8.4 astronomical units from the Sun, and passed through superior conjunction. The alignment between the spacecraft, the Sun and the Earth was so good that the tracking signal passed within little more than half a solar radius of the Sun's surface, leading to a Shapiro delay of over 260 microseconds. Tracking data was taken regularly for 15 days on either side of conjunction. Cassini was so far from the Sun that the buffeting effects of the solar wind and radiation pressure were negligible, so "anchoring" to a planet was not needed. Tracking the spacecraft was done using radar signals at two frequencies, X-band (7,175 megahertz) and Ka-band (34,316 megahertz), which made it possible to account for the small delay induced by the passage of the signals through the ionized corona of the Sun. This effect depends on the frequency of the signal, whereas the Shapiro delay does not. Twenty-five years of advances in transponders, atomic clocks and computational capabilities didn't hurt.

This "perfect storm" of happy chance produced a test of Einstein's delay effect that has not been surpassed. Since 2002 there have been numerous missions involving planetary orbiters, including Mars and Venus Express, Mars Reconnaissance Orbiter and Mercury MESSENGER, yet none has been able to improve upon Cassini's result.

Gravity's effect on light has by now been so thoroughly tested and confirmed that it has become useful to assume that general relativity is correct, at least with regard to this effect, and to use the bending and delay of light as a tool to explore other phenomena.

The classic example of using Einstein's theory as a tool for something else is the "gravitational lens." In 1979, astronomers Dennis Walsh,

Robert Carswell and Ray Weymann, using telescopes of the University of Arizona and Kitt Peak National Observatory, discovered a system that they initially called the "double quasar." This system, listed in astronomical catalogues as Q0957+561, was a pair of quasars separated in the sky by about 6 arcseconds. This by itself would not have been so unusual were it not for the fact that the two quasars were uncannily similar: their velocities of recession from the Earth were identical, within the precision of the measurement, and their spectra were almost identical. The only apparent difference was that one member of the pair was somewhat fainter than the other. The astronomers who discovered this system immediately proposed an explanation. They argued that there was actually only one quasar and that somewhere along the line of sight between us and it was a massive object that was deflecting the light from the quasar in such a way as to produce the multiple images (see Figure 3.10). The subsequent detection of a faint galaxy between the two quasar images along with a surrounding cluster of galaxies confirmed this interpretation. Since that time, the gravitational lens has become an important tool for astronomers and cosmologists.

The idea that a massive object could produce an image by gravitational lensing was not new. Ironically, Einstein was probably the first to consider the possibility of a gravitational lens, although he didn't publish it.

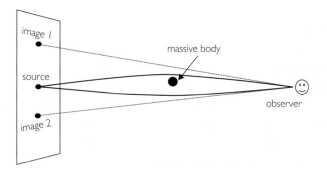

Fig 3.10 Gravitational lens. Two light rays emitted by a source pass by a massive body. Each is deflected by the curved spacetime around the body, the ray passing closer deflected a little more than the ray passing farther away. The observer sees the two rays coming from the direction of image 1 and image 2.

Indeed, the fact that he did this calculation was unearthed only in 1997. In the course of studying Einstein's original notebooks, science historian Jürgen Renn and colleagues came across a notebook from around 1912 in which Einstein worked out the basic equations for gravitational lenses, including the possibility of double images for a lens consisting of a simple massive body, and he derived a formula for the magnification of each image. Everything he did was off by a factor of two, of course, because he was using his 1911 formula for the deflection of light based on the principle of equivalence. In his notebook, he remarked that the effects were too small and that the probability of two stars being so perfectly aligned one in front of the other to produce an astronomical lens was too low ever to be of interest, so he didn't publish the calculations.

The physicist Oliver Lodge suggested the possibility of lenses shortly after the eclipse confirmations of the deflection of starlight in 1919, and in 1924 Orest Chwolson pointed out that if the source was perfectly aligned directly behind the lensing star, the image would be a perfect ring, today called the Einstein ring. In 1936, Einstein published a short note on gravitational lenses based on his earlier notes, but now with the correct factor. Apparently he did this primarily to get a retired engineer named Rudi Mandl to stop nagging him about it. He wrote to the editors of *Science* that "Mandl squeezed it out of me . . . , but it makes the poor guy happy."

But in 1937 the astronomer Fritz Zwicky pointed out that galaxies or even clusters of galaxies could act as gravitational lenses, thus relaxing the need for such precise alignment between the source, the lens and the Earth. The large mass associated with a galaxy can provide plenty of warpage of spacetime to deflect light rays, but because galaxies are mostly empty space (and clusters of galaxies even more so), light can easily pass through them, just as light passes through a glass lens.

The actual discovery of gravitational lenses gave general relativity a new astronomical role. For example, the number of quasar images, their relative brightness and placement, and any distortion in their shape all depend in detail on the distribution of matter in the intervening galaxy or cluster of galaxies. This is especially important because it is now widely believed that galaxies and clusters are embedded in halos of "dark matter," and that the mass of these dark halos can be anywhere between 10 and 100 times the mass of the visible galaxy or cluster. Even though this matter evidently does not produce light, its mass can warp spacetime and bend any light that goes through it. Thus, gravitational lensing is playing a major role in mapping the distribution of dark matter in the universe.

In 2003 a planetary system outside our own solar system was discovered using gravitational lensing, adding to the ever growing list of "exoplanet" systems discovered by other techniques, such as detecting the wobble of the star caused by its orbital motion relative to its planets. In this case, the combined lensing of a distant source by a star and its Jupiter-scale planet was measured and could be deconvolved to determine the ratio of the two masses and the approximate distance between the star and its planet. Additional systems were discovered subsequently, and gravitational lensing is proving to be a useful tool in the search for exoplanets.

Ilse Rosenthal-Schneider, one of Einstein's students in 1919, was amazed at his remarkably serene reaction to the telegram from Eddington announcing the eclipse results. When she asked how he would have felt if the observations had not confirmed his prediction, he answered: "Then I would have been sorry for the dear Lord. The theory is correct." Einstein was joking, of course. He understood full well that a theory stands or falls on the basis of its agreement with experimentation. Yet to his mind, general relativity was so beautiful, so elegant, so internally consistent that it had to be correct. The eclipse results merely justified his already supreme confidence. In their various ways, Don Bruns, VLBI radio astronomers and the Cassini spacecraft have shown that, so far, Einstein's confidence was well placed.

CHAPTER 4

Does Gravity Do the Twist?

Gravity Probe-B, the Relativity Gyroscope Experiment, may go down in the history of physics as one of the most difficult, most costly and longest physics experiments ever performed. From conception to completion it took almost half a century and cost $750 million, while the actual data taking took only sixteen months. The experiment was the brainchild of three naked men basking in the noonday California sun in the closing weeks of 1959. The three were all professors at Stanford University in Palo Alto. One of them was the eminent theoretical physicist Leonard I. Schiff, well known for his pioneering work in quantum theory and nuclear physics. In the late 1950s, however, he had become interested in gravitation theory. The second professor was William M. Fairbank, an authority on low-temperature physics and superconductivity, who had just arrived at Stanford in September of 1959, lured there from Duke University in North Carolina. The third was Robert H. Cannon, also a recent acquisition by Stanford, an expert in aeronautics and astronautics from MIT.

But before we learn how these naked professors came to formulate this experiment, let us first answer the question, what does a gyroscope have to do with relativity? When we think of a gyroscope we imagine something like a spinning flywheel. If the flywheel spins rapidly enough, its axis of rotation always points in the same direction, no matter how we rotate the platform or laboratory in which it sits, as long as the gyroscope is mounted on the platform using gimbals that allow it to turn freely with minimum friction. In other words, the axis of the gyro always points in a fixed direction relative to inertial space or to the distant stars. The difficulty you have in turning a rapidly spinning bicycle wheel is an everyday example of this gyroscopic effect. This, of course, is the basic principle behind the use of gyroscopes in navigation of ships, airplanes, missiles and spacecraft (GPS has now taken over many aspects of such navigation, of course). When attached more rigidly to a platform, this gyroscopic action is what keeps personal transporters like Segways or hoverboards from toppling over. However, according to general relativity, a gyroscope moving through curved spacetime near a massive body such as the Earth will not necessarily point toward a fixed direction; instead, its axis of spin will change slightly, or *precess*. Two distinct general relativistic effects can cause such a precession.

The first of these is called the "geodetic effect," and is a consequence of curved spacetime. Our everyday experience with gyroscopes tells us that as a gyroscope moves through space, its spin axis should maintain the same direction, a direction parallel to its previous direction. However, in curved spacetime, parallel in the local sense does not necessarily mean parallel in the global sense, and so upon completing a closed path, the gyroscope axis can actually end up pointing in a different direction than the one it started with.

A simple way to see how this can happen is to imagine a two-dimensional world, much like that of the nineteenth-century book *Flatland* by E. A. Abbott, but here confined to the surface of a sphere. Because the inhabitants of this "Sphereland" are only two-dimensional, they can't really construct the right kind of gyroscopes; as an alternative, they can take a little pointer and slide it about on their sphere in a way that keeps it always parallel to its previous direction, veering neither to the right nor to the left (up and down are not options in Sphereland). The pointer's tip then plays a role analogous to the spin axis of our gyroscope. To demonstrate what can happen, the Spherelanders consider the following closed route (see Figure 4.1): from a point at zero degrees longitude on the equator, move east along the equator to 90 degrees longitude, then

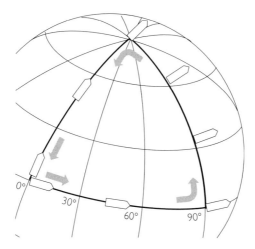

Fig 4.1 Precession in Sphereland. A pointer is carried parallel to itself from 0 degrees longitude to 90 degrees longitude. The path then turns north, but the pointer continues to point east, and maintains that direction up to the North Pole. A right-angle turn of the path leaves the pointer pointing to the rear. The pointer continues to point to the rear (north) back to the starting point. The result is a precession of the direction of the pointer from an easterly direction to a northerly direction.

go due north to the North Pole, make a 90 degree turn, and go due south back to the starting point on the equator. Suppose the Spherelanders start the pointer out parallel to the equator, pointing east. When they reach the first turn the pointer will still point east, and when they head north the pointer will now be perpendicular to the path. At the North Pole they make a 90 degree left turn, but now the pointer's direction is to their rear, which is to the north as they head south. When they reach the equator once again, the pointer, which has been kept parallel to itself all the way, now points north, whereas it started out pointing east. This change in direction of the pointer is what we would call precession in the case of a gyroscope. The curvature of the two-dimensional surface of Sphereland accounts for this precession, and we understand it without much difficulty. The difference between this example and the geodetic effect on a moving gyroscope is that it is the curvature of spacetime and not just the curvature of space that is important.

The geodetic effect has been known since the early days of general relativity. The first to calculate the effect was Willem de Sitter, the Dutch theorist who had played a pivotal role in bringing general relativity to the attention of Eddington and the British physics community. In a paper published in the *Monthly Notices of the Royal Astronomical Society* less than a year after Einstein's November 1915 papers on general relativity, de Sitter showed that relativistic effects would cause the axis perpendicular to the orbital plane of the Earth–Moon system to precess at a rate of about 0.02 arcseconds per year. De Sitter was not thinking in terms of gyroscopes; instead, he had in mind how the combined relativistic gravitational fields of the Earth and Sun would perturb the Earth–Moon orbit. However, Eddington and others soon pointed out that the Earth–Moon system is really a kind of gyroscope, with the axis perpendicular to the orbital plane playing the role of the gyroscope's rotation axis, so the de Sitter effect was effectively a precession of the Earth–Moon gyroscope. But if this is the case, then the Earth, as it spins about its own rotation axis, is also a gyroscope, so in fact both the Earth and the Earth–Moon system should precess in the same way. At the time, measuring such a small effect was hopeless. Only in recent years has a technique called "lunar laser ranging" (see Chapter 5 for a discussion) given such precise information about the Earth–Moon orbit that the de Sitter effect could be measured, to around half a percent.

Instead of the Earth–Moon system, consider a more down-to-Earth situation: a laboratory-size gyroscope orbiting the Earth with its axis lying in the orbital plane, say pointing vertically (see the gyroscope labeled "start" in Figure 4.2). Without general relativity, the gyroscope would maintain its direction relative to distant stars as it orbits the Earth, and so after a complete orbit it would again be pointing in the same vertical direction. But general relativity predicts that as the orbit carries the gyroscope around through the Earth's curved spacetime, the gyroscope will experience a precession within the orbital plane at a rate of a little over a thousandth of an arcsecond per orbit. The direction of the precession is in the same sense as the motion of the gyroscope around its path, counterclockwise if looking down on the orbit from above (see Figure 4.2). Since the period of revolution of a low Earth

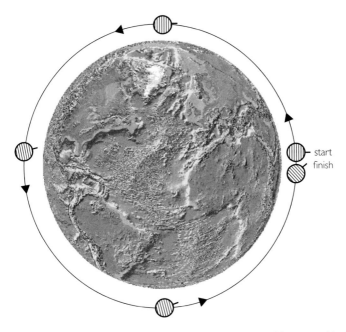

Fig 4.2 Geodetic precession of a gyroscope in near-Earth orbit. After one orbit, the direction of the gyroscope axis has rotated relative to its initial direction in the same sense (counterclockwise) as that of the orbit. The net effect over one year (5,000 orbits) is 6 arcseconds.

orbit is about 1.5 hours, the net precession in one year will be around 6 arcseconds. This is the geodetic effect.

The other important relativistic effect on a gyroscope is known as the dragging of inertial frames, one of the most interesting and unusual of the predictions of general relativity (see Figure 4.3). The origin of this effect is the rotation of the body in whose gravitational field the gyroscope resides. According to general relativity, a rotating body attempts to "drag" the spacetime surrounding it into rotation. The simplest way to get a picture of the consequences of this dragging is to use a fluid analogy.

Consider a large and deep swimming pool with a very large drain in its center. Water flows down this drain, producing a whirlpool at the surface of the kind we commonly see in bathtubs. To keep the level of the pool constant, let us assume that the water lost down the drain is continuously

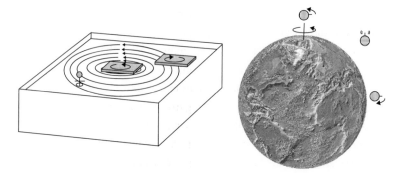

Fig 4.3 Left: Swimmer and air mattresses in a pool with a whirlpool. All are tethered to the bottom of the pool so that they don't move around the pool. The air mattress at the edge of the whirlpool rotates in a clockwise direction because water closer to the center moves faster, while the air mattress at the center rotates counterclockwise with the water. A vertical swimmer treading water does not rotate. Right: Dragging of inertial frames. Stationary gyroscopes near a rotating Earth can precess because of dragging of spacetime by rotation of the Earth. If the axis lies perpendicular to the rotation axis of the Earth, the precession will be opposite to the Earth's rotation for a gyroscope at the equator, and with the Earth's rotation for a gyroscope at the pole. If the axis is parallel to the Earth's rotation axis then there is no precession. For other locations and other orientations of the gyroscope axis, the precession will be between these extremes.

being replaced through inlets at the sides of the pool. Imagine now three Stanford professors floating in the pool. Professor Schiff is floating on an air mattress between the whirlpool and the edge of the pool, with his feet closest to the drain. Professor Fairbank is floating on a similar air mattress but is straddling the whirlpool, well above the drain of the deep pool. Professor Cannon is treading water. For simplicity's sake, let us also assume that each professor is anchored to the bottom of the pool by a tether attached to his waist. This is to prevent them from circling the drain, which would complicate the effect we are looking for. With this arrangement, the behavior of these professors is very similar to that of three gyroscopes in a spacetime being dragged by a rotating body. As with all the analogies for relativistic effects that we have used in this book, one must be careful not to push the analogy too far. Gyroscopes in spacetime are not the same as air mattresses in water, but if the analogy helps us remember the qualitative effects, it is a useful one.

First consider Professor Schiff. Because the water closer to the whirlpool moves around more quickly than the water farther away, the foot of his air mattress is dragged more quickly than the head, and so while the whirlpool rotates, say, counterclockwise as seen from above, Schiff's air mattress rotates or precesses in a clockwise direction. This is precisely the behavior of the axis of a gyroscope on the equatorial plane in a dragged spacetime, with its axis pointing outward (Figure 4.3). Contrast this behavior with that of Professor Fairbank, whose air mattress straddles the whirlpool. The head and foot of his mattress are also pulled by the water, but because they are on opposite sides of the whirlpool the mattress is pulled in the same sense as the whirlpool, in other words counterclockwise. This is just what happens to a gyroscope on the rotation axis of the dragged spacetime, with its own axis perpendicular to the rotation axis. Finally, we see that Professor Cannon, who is treading water, doesn't do much of anything. The direction of his body remains vertical, no matter where he goes in the pool. The same is true for a gyroscope whose axis is parallel to the rotation axis of the central body; the dragging of spacetime has no effect on it.

Of course, there is another key difference between the air mattress precession and the gyroscope precession due to the dragging of inertial frames: size. The predicted precession for a gyroscope on the equator of the Earth is only one-tenth of an arcsecond per year. Unlike the geodetic precession, the dragging of inertial frames does not depend on whether or not the gyroscope is moving through spacetime (the air mattresses precessed even though they were stationary in the pool), and so there is little difference in this case between the precession of a gyroscope on the Earth and that of a gyroscope in orbit. For a low Earth orbit it is between 0.1 and 0.05 arcseconds per year, depending on the tilt of the orbit relative to the Earth's equatorial plane and on the initial direction of the gyroscope axis relative to the Earth's rotation axis.

What makes this effect so interesting and important is that while the other effects that we have described in this book, including geodetic precession, have to do with such concepts as gravitational fields, curved spacetime, and nonlinear gravity, this effect tells us something about the inertial properties of spacetime. If you ask yourself "Am I rotating?" and you wish an answer with more accuracy than you can get simply by seeing if you are getting dizzy, you usually turn to a gyroscope, for the axis of a gyroscope is assumed to be non-rotating relative to inertial space. If you were to build a laboratory whose walls were constructed to be lined up with the axes of three gyroscopes arranged to be perpendicular to each other, you would conclude that your laboratory was truly inertial (and if the laboratory were in free fall, that would be even better). However, if your laboratory happened to be situated outside a rotating body, the gyroscopes would rotate relative to the distant stars because of the dragging effect just described. Therefore, your laboratory can be non-rotating relative to gyroscopes, yet still rotate relative to the stars. Relativists make a careful distinction between a laboratory that is *locally* non-rotating, that is tied to gyroscopes, and one that might rotate relative to distant stars. In this way, general relativity rejects the idea of absolute rotation or absolute non-rotation, just as special relativity rejected the idea of an absolute state of rest.

To understand this more clearly, contrast it with Newtonian theory. True, Newtonian theory proposed that all inertial frames are equivalent, regardless of their state of motion, but it still had to allow an absolute concept when it came to rotation. A simple example, known as Newton's bucket, will illustrate this idea (see Figure 4.4). Fill a bucket with water, place it on a turntable, and start the turntable spinning. As the bucket starts to spin, the water doesn't do much of anything at first, but eventually the friction between the water and the walls of the bucket causes the water to spin along with the bucket. As a consequence of this, the surface of the water becomes concave and the water begins to climb up the sides of the bucket; a depression forms in the center. Quite naturally, we attribute this behavior to centrifugal forces pushing the water away from the rotation axis. When the turntable is stopped, eventually the water slows down and returns to its initial state with a flat surface.

As mundane and commonplace as this simple observation is, it has led to some of the most intriguing and vexing philosophical questions. Newton himself wrestled with them. One question is, how does the water know that it is rotating and should have a concave surface instead of a flat one? If we truly abhor the concept of absolute space, as relativity in either its Newtonian or Einsteinian forms teaches us, we cannot answer that the water knows that it is rotating relative to absolute, non-rotating space. With respect to what then? The best we can do is to answer that somehow the water knows that it is rotating relative to the distant stars and galaxies. As reasonable as this sounds, it does beg two questions. Suppose we performed this bucket experiment in an otherwise completely empty universe. With nothing to which to refer its state of motion, would the water know what to do as the turntable spun? Would its surface become concave or stay flat? That is the first question, to which there is no satisfactory answer based on physics. Up to a point, of course, this question is irrelevant, because we don't live in an empty universe anyway.

The second question is somewhat more meaningful: Suppose we leave the bucket at rest, and let the entire universe rotate around it with the same rotation rate as the bucket had in the previous experiment, but in the opposite sense. Would the water become concave as before? If only the rotation of the bucket relative to the distant matter in the universe is important, then the two experiments should give the same concave shape for the water's surface. In other words, it should not matter whether we say that the universe is non-rotating and the bucket is rotating, or that the bucket is non-rotating and the universe is rotating. Only the rotation of one relative to the other is relevant.

Unfortunately, Newtonian gravity predicted that the rotating universe would have no effect on the bucket, and therefore you *had* to invoke an absolute space to understand rotation. But in general relativity, the dragging of inertial frames provides the way out of this absolutism. As early as 1923, Eddington suggested as much in his beautiful textbook on general relativity. However, it wasn't until the mid 1960s that theorists could show that the dragging effect provides a good accounting of how rotation is indeed relative. The demonstration consisted of a simple model calculation of the following situation: Imagine a spherical shell of matter, like a balloon, that is rotating about some axis (for the purposes of this discussion we can ignore the flattening of the balloon caused by centrifugal forces). At the center of the shell is a gyroscope with its spin axis perpendicular to the axis of rotation of the balloon. According

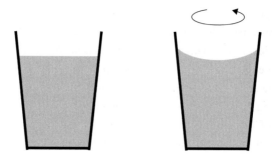

Fig 4.4 Newton's bucket. Left: The bucket is not rotating and the surface of the water is flat. Right: The bucket is rotating and the surface of the water is concave. Would the surface be concave if the bucket were "not rotating" and the universe rotated around it?

to Newtonian gravitation, the interior of the balloon is absolutely free of gravitational fields. The gyroscope feels no force whatsoever. To a first approximation, the same is true in general relativity, except for the dragging of inertial frames effect, which produces forces in the interior of a rotating shell just as it would in the exterior. The effect of these forces is to cause the gyroscope to precess in the same direction as the rotation of the shell, but as you might imagine from our previous discussion, for a shell of planetary dimensions, say of the radius of a typical planet and containing the mass of a typical planet, the rate of precession is very small, much smaller than the rate of rotation of the shell.

But now imagine increasing the mass of the shell and increasing its radius (keeping its rate of rotation the same), and consider the limit in which the mass tends toward the mass of the visible universe and the radius tends toward the radius of the visible universe. The remarkable result is that as you increase these values, the rate of precession of the gyroscope in the center grows and, in the limit, tends toward the rate of rotation of the shell. In other words, inside a rotating universe the axes of gyroscopes rotate in step with the rotation of the universe; their axes are tied to the directions of distant bodies in that universe. Therefore, a laboratory tied to the gyroscopes, which we would define to be non-rotating, would indeed be non-rotating relative to the distant galaxies. Imagine now that we placed a bucket inside the shell instead of a gyroscope, and imagine that we kept the bucket fixed, or non-rotating, with the shell rotating around it. As you expand the size of the rotating shell, residual frame dragging forces would cause the water to climb the sides of the bucket. Therefore, an observer in this scenario would see exactly the same physical phenomenon as would an observer looking at a rotating bucket inside a non-rotating universe. The existence of the dragging of inertial frames then guarantees that rotation must be defined relative to distant matter, not relative to some absolute space. This is what makes the detection of this effect so vital.

In addition to resolving this conceptual problem, the relativistic frame dragging effect has important astrophysical implications beyond the solar system. Astronomers have found that some of the incredible out-pouring of energy from quasars is directed along narrow jets of matter that stream at nearly the speed of light in opposite directions from a compact central region. The leading model for this phenomenon involves a vast disk of hot gas spiraling inward around a spinning supermassive black hole (we will return to these beasts, including the one in our own Milky Way, in Chapter 6). The combination of the magnetic fields generated by the charged particles in the gas and the extreme dragging of spacetime in the vicinity of the black hole generates strong electric fields that can accelerate particles away from the black hole along its spin axis. The particles can reach speeds close to that of light, and upon interacting with the twisted magnetic field lines they can emit radio waves and other forms of electromagnetic radiation. These radio jets can be seen extending millions of light years away from the central quasar. Rotating black holes represent extreme examples of the effect of the dragging of inertial frames, and so it would be very desirable to verify that this effect exists.

All well and good. But still, the effects on gyroscopes on and near the Earth are horribly small. What would possess anyone to actually try to measure them? It is here that the three Stanford professors return to the story.

At Stanford University in the 1950s, back before the days of coeducational athletic facilities, the Encina gymnasium and its walled-in, open-air swimming pool was restricted to males only (the women's gym was on the other side of the campus). As such, it was customary for users to swim in the nude. Schiff had a virtually unshakable routine of going to the Encina pool every day at noon, swimming 400 yards, and eating a bag lunch afterwards while sunbathing. Even though he was chairman of the physics department, he would try his best to schedule meetings and appointments so as not to conflict with his noon swim. Fairbank knew about Schiff's daily routine, and when he bumped into Cannon on campus one day in late 1959 and they began to talk about gyroscopes, Fairbank suggested that they go see Schiff at the swimming pool.

Each of these men had had gyroscopes on his mind for a while. Schiff had been thinking about gyroscopes ever since he opened his December 1959 issue of the professional physicists' magazine *Physics Today* and saw the advertisement on page 29. There, hovering in an artist's conception of interstellar space, was a perfect sphere girdled by a coil of electrical wires, captioned "The Cryogenic Gyro." The advertisement announced the development at the Jet Propulsion Laboratory in Pasadena of a super new gyroscope consisting of a superconducting sphere supported by a magnetic field (from the coils), all designed to operate at 4 degrees above absolute zero. Schiff had taken a strong interest in tests of general relativity lately, and so he asked himself whether such a device could detect interesting relativistic effects. During the first two weeks of December he carried out the calculations, finding both the well-known geodetic effect, as well as the dragging of inertial frames effect. The latter discovery was entirely new, at least as applied to gyroscopes. Back in 1918, two German theorists, Josef Lense and Hans Thirring, had shown that the rotation of a central body such as the Sun would produce frame dragging effects on planetary orbits that were unfortunately utterly unmeasurable at the time, but no one had apparently looked at the effect of the rotation of a central body on gyroscopes.

Fairbank's field was low-temperature physics, the properties of liquid helium, and the phenomenon of superconductivity, the disappearance of electrical resistance in many materials at low temperatures. He had also been thinking about the potential for a superconducting gyroscope that could be built in the new laboratory that he was setting up at Stanford, and he and Schiff had begun to talk about how these relativistic effects could be detected. Fairbank suggested measuring the precessions using gyroscopes in a laboratory on the equator, but this did not look promising. The reason was gravity. The best gyroscopes of the day had as their main element a spinning sphere, just as in the JPL advertisement. But the sphere had to be supported against the force of gravity, and the standard method of doing this was by electric fields or by air jets. Unfortunately, the forces required to offset gravity were so large that they introduced spurious forces or torques on the spinning ball that gave its spin axis a precession thousands of times larger than the relativistic effect being sought after, though easily small enough to permit accurate navigation and other commercial uses. This problem would effectively go away if the gyroscope were in orbit, where the gravitational forces are zero to high accuracy, and essentially no support is required. But remember, this was only two years after the Soviet Union's launch of Sputnik, the first orbiting satellite, and Schiff and Fairbank could not imagine realistically being able to do this.

This was where Cannon came in. Cannon knew gyroscopes. He had helped develop gyroscopes used to navigate nuclear submarines under the Arctic icecap. He also knew aeronautics, and he was active in the fledgling space race that Sputnik had started. Before coming to Stanford from MIT, Cannon had already begun to consider the improvements in spacecraft performance that would come with orbiting gyroscopes.

Finally, the three were together (in their birthday suits) at the Stanford pool. When Schiff and Fairbank told Cannon about the proposed experiment, Cannon's first response was astonishment. To pull it off, they would need a gyroscope a million times better than anything that existed at that time. His next response was: Forget about doing it on Earth, put it into space! An orbiting laboratory is not at all farfetched, and in fact NASA was already laying plans for an orbiting astronomical observatory. Furthermore, Cannon knew the right people at NASA whom they could contact. With that, a five-decade adventure had begun. Only Cannon would live to witness the end of the story.

It is one of those strange twists of scientific history that, almost simultaneously with Schiff, Fairbank and Cannon, someone else was thinking about gyroscopes and relativity. Completely independently of the Stanford group, George E. Pugh at the US Pentagon was doing the same calculations. Pugh worked for a section of the Pentagon known as the Weapons Systems Evaluation Group, and for him, toying with gyroscopes was a perfectly reasonable activity because gyroscopes have obvious military applications in the guidance of aircraft and missiles. In a remarkable memorandum dated 12 November 1959, Pugh outlined the nature of the two relativistic effects, although he had the frame dragging effect wrong by a factor of two, and described the requirements for detecting them using an orbiting satellite. Some of Pugh's ideas, such as a technique for compensating for the atmospheric drag felt by the satellite, ultimately became important ingredients in the Stanford experiment. It is highly unlikely, however, that the Pentagon actually incorporated relativistic gyroscope effects into its military guidance systems. Pugh's classified work could not be published in the open scientific literature, and so Schiff was initially given credit for the idea of a gyroscope test. Only later was Pugh's work declassified and recognized for equal credit.

In January 1961, Fairbank and Schiff kicked off the experiment officially with a proposal to NASA for an orbiting gyroscope experiment. Fairbank also recruited a young low-temperature physicist named C. W. Francis Everitt to join the Stanford project. Born in England in 1934, Everitt had received a Ph.D. from Imperial College in London in 1959, followed by a post-doctoral stint at the University of Pennsylvania, working on liquid helium. In late 1963, NASA began funding the initial research and development work at Stanford to identify the new technologies that would be needed to make such a difficult measurement possible. In 1971, NASA selected its Marshall Space Flight Center in Huntsville Alabama as program manager, both for the Stanford experiment and for the rocket redshift experiment that was being developed by Robert Vessot at Harvard. NASA headquarters designated Vessot's experiment as Gravity Probe-A, planned for a 1976 launch, to be followed soon thereafter by the gyroscope experiment, designated Gravity Probe-B or GP-B. While the redshift experiment went off as planned (see Chapter 2), things turned out rather differently for GP-B. It is not known if serious plans were ever made for a Gravity Probe-C, Gravity Probe-D, and so on.

Stanford's modest research and development effort lasted until about 1981, when Everitt became Principal Investigator of the project, and soon it moved toward the mission design phase. At that time, plans called for a preliminary flight on board the Space Shuttle to test key technologies to be used in GP-B, followed by a launch of the full spacecraft from the Shuttle a few years later. Unfortunately the 1986 Space Shuttle *Challenger* catastrophe forced a cancelation of the technology test, and a complete redesign of the spacecraft for a launch from a Delta rocket.

The goal of the experiment was to measure both the geodetic effect and the frame dragging effect to an accuracy of better than a milliarcsecond per year. Because the smaller frame dragging effect is only about 40 milliarcseconds per year for the orbit being planned (a polar orbit at an altitude of about 642 kilometers), this meant that a one or two percent measurement of this effect would be possible. The task of building an orbiting gyroscope laboratory that could measure such tiny effects put the Stanford scientists at or beyond the frontiers of experimental physics and precision fabrication techniques, presenting them with apparently insuperable problems. Miraculously, they managed to overcome each one.

A brief description of the experiment will illustrate the things that had to be done. The gyroscopes (for redundancy, there were four) were spherical rotors of fused silica, about 4 centimeters in diameter, housed in a chamber (see Figure 4.5). The rotors had to be uniform in density and perfectly spherical in shape to better than one part in a million. This is like imagining an Earth where the tallest mountain and deepest valley reach 1 meter! The reason for this requirement is that gravitational forces from the Earth and Moon and from the spacecraft itself would interact with any mass irregularities in the rotor and cause spurious precessions. Similar effects caused by gravitational forces from the Sun and Moon acting on the Earth's equatorial bulge make the Earth's rotation axis precess with a period of about 26,000 years, causing the North Star to appear to wander from true north. Overcoming the problems of making a perfect sphere and then testing how spherical it is to the above precision required the invention of completely new fabrication and testing procedures.

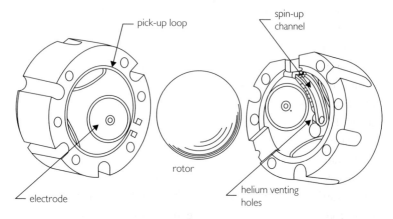

Fig 4.5 GP-B rotor and its enclosing chamber. In space, the rotor is levitated inside the spherical chamber. Six electrodes spaced around the wall of the chamber sense if the rotor gets too close and send signals to the spacecraft to nudge it in the proper direction. The "pick-up loop" is a superconducting wire embedded in the wall to measure any changes in direction of the magnetic field of the rotor caused by precession of its spin. The "spin-up channel" lets helium gas from the surrounding dewar of liquid helium flow past the rotor, spinning it up to around 4,000 revolutions per minute, after which the gas is vented to space.
Credit: NASA and Stanford University.

The main reason for going into space was to avoid having to support the gyroscopes against the force of gravity, because those support forces can generate spurious precessions. Unfortunately, while the gyroscopes move in complete free fall inside the spacecraft, the spacecraft itself is being pushed around by the residual atmosphere of the Earth, by the solar wind and by periodic attitude control forces required to orient the spacecraft. How do you avoid having the gyroscopes collide with the walls of the spherical chambers inside which they are spinning? The answer is a technique called "drag-free control." Six circular electrodes are installed in the walls of the spherical chambers in which the rotors reside, so that if a rotor gets too close to an electrode a signal is sent to thrusters that nudge the spacecraft a little bit to keep the separation at a pre-selected value. This was a delicate and critical technology, because the average gap between each rotor and the wall of its spherical chamber was about one-thirtieth of a millimeter. We will return to drag-free control in Chapter 9, when we discuss LISA, the planned gravitational wave detector in space.

If a rotor is perfectly spherical, how do you determine the direction of its spin? You can't just attach a stick to the rotor at one of the poles, because the stray gravitational forces acting on the mass of the stick would cause enormous precessions that would swamp the relativity effects. The solution was to coat each rotor with a thin, perfectly uniform layer of the element niobium. When the ball is spinning at low temperatures, near absolute zero, the niobium becomes a superconductor, its electrical resistance vanishes, and it develops a magnetic field whose north and south poles are exactly aligned with the rotation axis of the rotor. A tiny superconducting wire, called a "pick-up loop," is embedded in the wall of the chamber that houses the rotor. If the axis of the rotor changes direction, the change in the magnetic field induces currents in the pick-up loop that are measured by very precise devices known as superconducting quantum interference devices, or SQUIDs, also operating at near absolute zero. This required new techniques for working near absolute zero using liquid helium, and adapting those techniques to a space environment. The spacecraft itself contained a very special "thermos bottle" or dewar to hold the 2,400 liters (over 600 gallons) of liquid helium and to maintain it at 1.8 degrees above absolute zero.

If the balls are perfectly spherical, how were they to be set spinning? The solution to this problem was to incorporate into the wall of the chamber that housed each rotor a small "spin-up channel" that forces helium gas past the sphere, using friction to get it spinning. The helium gas came from natural "boil off" from the liquid helium used to cool the apparatus (no thermos can keep liquid helium cold enough to not boil some of it). At the start of the space flight the four rotors were spun up to around four thousand revolutions per minute, after which venting holes in the housing chamber allowed the helium gas to escape to the vacuum of space.

As we described previously, the gyroscopes precess relative to the distant stars, so a very accurate telescope had to be designed and built into the spacecraft package to determine a reference direction accurate to the milliarcsecond level per year. The spacecraft was controlled so that the telescope always pointed toward a selected star, called IM Pegasi. This star lies at a distance of about 300 light years from Earth, about 17 degrees north of the equator. In addition to being optically bright and relatively isolated in the sky, it was also bright in the radio band. This was important because, being in the environment of the Milky Way, it moves, and thus its own motion relative to truly distant objects, i.e. the quasars, needed to be measured to the required precision using VLBI (see Chapter 3).

While simple to state in words, each of these problems was a major multi-year research and fabrication project, and integrating all the components into a functioning spacecraft was a major challenge. The cancelation of the Shuttle test mission in 1986 and the spacecraft redesign resulted in delays and cost overruns for the GP-B program. Similar delays and budget problems with the Hubble Space Telescope, combined with the discovery of its flawed main mirror following its launch in 1990, caused considerable anxiety at NASA about budgets, and worries among astronomers and space scientists about funding for their own projects. One result was rising criticism of the GP-B program and calls for its cancelation. In fact, on more than one occasion NASA and the Office of Management and Budget, the fiscal oversight arm of the US Administration, would set the GP-B budget to zero for the subsequent fiscal year, effectively canceling the project, only to find that members of Congress would restore the budget following judicious lobbying by Francis Everitt and other supporters of GP-B.

In 1992, Daniel Goldin was appointed NASA Administrator by President George H. W. Bush, and he was determined to end the bickering over GP-B. He asked the National Academy of Sciences to conduct a thorough review of the project, promising to abide by their recommendation, whether it be thumbs up or thumbs down.[1] In addition to investigating the technical challenges that remained to be overcome

---

[1] Disclosure: Cliff was a member of the National Academy panel, and later was appointed by NASA as Chair of an external Science Advisory Committee for Gravity Probe-B from 1998 to 2011.

and estimating the remaining cost of the mission, the panel debated whether the scientific return of the mission was worth it. This debate was not trivial.

When GP-B was first conceived in the early 1960s, tests of general relativity were few and far between, and most were of limited precision. But by 1994 there had been enormous progress in experimental gravity in the solar system and in binary pulsars, as we have described in this book. Some panel members argued that the many experiments had so constrained the theoretical possibilities in favor of general relativity that GP-B would not give any improvement or new information. The counter-argument was that all the prior experiments involved phenomena entirely different from the precession of a gyroscope, and therefore that GP-B was testing something potentially new. Another issue was that, if GP-B were to give a result in disagreement with general relativity, it would very likely not be believed, and given the high cost of the experiment, the probability of repeating it was extremely small. In the end, while the panel was not unanimous, a majority did recommend going ahead with GP-B, and Goldin committed NASA to the project. NASA then engaged the aerospace company Lockheed Martin in Palo Alto to build, integrate and test the spacecraft in collaboration with Stanford and Marshall Space Flight Center.

The satellite finally was launched on 20 April 2004, and injected into an almost perfectly circular polar orbit at an altitude of 642 kilometers above the Earth's surface, precisely as planned. Almost every aspect of the spacecraft, its subsystems and the science instrumentation performed extremely well, some far better than expected. The plan of the mission was to begin with a three-month period of testing, calibration and fine-tuning. This would be followed by 12 months of science data taking, and a final month of additional calibrations. Early on in the science phase, the data showed clearly the larger geodetic precession of all four gyroscopes, giving initial hope that all would go well. That hope was soon dashed when a nasty source of error reared its head. Each rotor appeared to be experiencing strange precessions of its spin axis, with no apparent pattern or commonality among them. As the sixteen-month mission period drew to a close in the fall of 2005, there was serious concern that the experiment would fail to detect the tiny frame dragging precession, the main goal of the mission.

What ensued during the data analysis phase following the mission was worthy of a detective novel. The critical clue came from the calibration tests carried out at the end of the mission. The four rotors had been set spinning with their spin axes initially parallel to the axis of the telescope directed to the guide star. The spacecraft also rolled slowly about this axis about once every 78 seconds. For one of the post-science tests, they deliberately forced the spacecraft to point away from the guide star by as much as 7 degrees. This was such an extreme maneuver that you would never try it at the start of the mission, because if anything went wrong the game would be over. After the data has been taken and safely stored away, it is worth the risk. As it happened, the rotors experienced unexpectedly large precessions of their spins during the maneuver, but there was a very specific pattern to the effects. Unraveling this pattern helped the GP-B team to determine that the extra precessions were being caused by interactions between random patches of *electric* potential fixed to the niobium surface of each rotor, and similar patches on the inner surface of its spherical chamber. Such patches were known to occur on superconducting niobium films, but pre-flight tests of the rotors had shown that the patches would be too weak to cause a problem. For some unknown reason, the rotors in the spacecraft developed stronger patches. This insight allowed the researchers to build a mathematical model for the "patch effects" and thereby to subtract the anomalous precessions from the data on each rotor. When this was done, all four rotors showed the same precession behavior, clearly revealing both the larger geodetic effect and the smaller frame dragging effect. The original goal of GP-B was to measure the frame dragging precession to about 1 percent, but the problems discovered over the course of the mission dashed the initial optimism that this was possible. Everitt and his team had to pay the price of the increase in measurement uncertainty that came from using a complex model to remove the anomalous precessions. The experiment uncertainty quoted in the final result was roughly 20 percent for the frame dragging effect, but the result agreed with general relativity.

This data analysis effort took five years. When the long-awaited results were finally announced at a NASA press conference on 4 May 2011, the feeling of many could be summed up by the opening line of the song by the great blues singer Etta James: "At laaaassst, my love has come along . . . " The half-century adventure started by three naked professors was over.

The story of Gravity Probe-B has all the ingredients of a case study in science politics, raising many thorny questions about how science, especially "big" science, is carried out. How do we balance the value of a scientific return against the cost of a project? What is the best way to make critical decisions, particularly concerning cancelation of projects? How do we weigh the value of different kinds of science, say fundamental physics versus astronomical discovery, in setting priorities or deciding among competing proposals or competing scientific constituencies?

For GP-B, these kinds of questions became even more relevant in 1986, when a young post-doctoral researcher at the University of Texas named Ignazio Ciufolini suggested a way to measure the frame dragging effect almost as accurately as the stated goal for GP-B, and at a tiny fraction of the cost. He pointed out that general relativity predicts that the tilted orbital plane of a body revolving around a rotating object such as the Earth will rotate by a small amount, in the same direction as the rotation, as a result of frame dragging. One consequence is that the point where the orbit crosses the equatorial plane of the Earth will rotate (see Figure 4.6). This was one of the effects that Lense and Thirring had pointed out in 1918. But in 1976 geophysicists had launched an Earth-orbiting satellite called LAGEOS, and Ciufolini realized that precision tracking of such satellites could potentially detect this frame dragging effect.

LAGEOS is an acronym for Laser Geodynamics Satellite, and it is about as simple a satellite as one could possibly imagine. It is a massive spherical ball of solid brass covered in aluminum weighing about 400 kilograms. The surface is studded with 426 fused silica glass mirrors, called retroreflectors, each in the shape of one corner of the interior of a cube (see Figure 4.6). A light ray that approaches any one of the "corner-cube" mirrors will bounce off one face then off an opposite face and then return in exactly the same direction from which it came. By

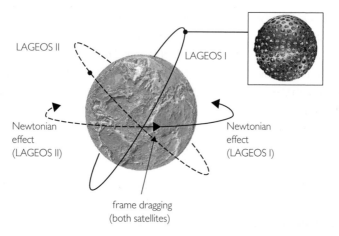

Fig 4.6 Frame dragging and LAGEOS. The rotation of the Earth causes the plane of an inclined LAGEOS orbit to rotate at about 30 milliarcsecond per year in the same sense as the rotation of the Earth (short arrow). The variations in the Newtonian gravity of the Earth caused by its flattening and by the uneven distribution of mass also cause the planes to rotate. The amount of rotation is as large as 126 degrees per year, but the direction depends on the inclination angle of the orbit. For LAGEOS I (solid arrow) and II (dashed arrow), the effects are in the opposite direction. Inset: The LAGEOS I satellite, showing the corner reflectors embedded on the surface. Credit: NASA

sending pulsed laser beams from Earth and measuring the round-trip travel time of the pulses, researchers can measure the distance between the laser and the satellite with sub-millimeter precision. This technique is called "laser ranging" and was developed during the late 1960s for precise ranging to the Moon (see page 138). The satellite orbits at an altitude of 5,990 kilometers (or 12,200 kilometers from the center of the Earth) on a nearly perfect circle. Because it is so massive and simple (for example, it has no large solar panels), the residual atmosphere at that altitude has almost no effect on it. It even contains a plaque designed by astronomer Carl Sagan providing information about the present Earth for future humanity (if any still exist) when the satellite reenters the atmosphere and falls to Earth in 8.4 million years! Because its orbit is so pure, geophysicists could use laser ranging to it to measure the shape of the Earth, and to study continental drift and other dynamics of the Earth's crust.

But Ciufolini immediately recognized a difficulty. The Earth is not perfectly spherical, but instead is slightly flattened at the poles and bulges a bit at the equator, a result of its rotation. The deviation is only one part in two thousand, but it modifies the Newtonian gravitational field of the Earth in such a manner that those deviations cause the plane of LAGEOS to rotate in the same direction as the relativistic frame dragging effect. Ciufolini calculated that the relativistic effect would be about 30 milliarcseconds per year. However, the effect due to the Earth's bulge is huge, 126 *degrees* per year, almost fifteen million times larger. There was no way to measure this tiny effect on top of such a large effect, and although the value of the Earth's flattening had been measured to reasonable accuracy, it wasn't good enough.

Nevertheless, Ciufolini had an idea to get around this problem. If there were a second LAGEOS satellite orbiting at the same altitude and in an equally circular orbit, but with its inclination relative to the equator chosen so that the two inclinations added up to 180 degrees, then the calculations showed that the Newtonian rotation of that orbit would be exactly the same as for the existing LAGEOS, but in the *opposite* sense. The relativistic rotation would be exactly the same size and in the *same* direction for both orbits; this effect does not depend on the tilt angle. Thus, one could measure the rate of rotation of the planes of each orbit and simply add them together. The Newtonian part would cancel exactly, leaving twice the relativistic part. Injecting a satellite into just the right orbit would be challenging, but doable. Ironically, Ciufolini's proposal was a variation of a 1976 idea by none other than Francis Everitt and his colleague Richard van Patten; their proposal was for two satellites in polar orbits about 600 kilometers above the Earth, moving in opposite directions. In this configuration the Newtonian effect is much, much smaller, making it easier to cancel and reveal the frame dragging precession. It never attracted much attention, and Everitt and van Patten returned their attention to Gravity Probe-B. Even though their proposal was published just six weeks before the launch of LAGEOS I, they were apparently unaware of what the geo-physicists were working on. Science is rife with missed opportunities caused by the compartmentalization of different fields. It would be ten years before Ciufolini would hit upon LAGEOS as a tool for measuring frame dragging.

At the time of Ciufolini's proposal, the geophysics community was planning to launch a second LAGEOS satellite in order to advance their studies of the Earth. Ciufolini and many relativists campaigned vigorously to have LAGEOS II launched with the special inclination angle of 70.16 degrees (LAGEOS I was at 109.84 degrees) needed to measure frame dragging, but other considerations prevailed in the end. LAGEOS II was launched in 1992 with an inclination of 52.64 degrees, mainly to optimize coverage by the world's network of laser tracking stations, which was important for geophysics and geodynamics research.

This was a major disappointment, but Ciufolini, now based in Italy, and his colleagues tried to make the best of it. Because the cancelation effect was not ideal, the errors in measuring the frame dragging rotation would be larger than they had hoped, but between 1997 and 2000 they reported having measured the 30 milliarcsecond effect to between 20 and 30 percent precision, although some critics argued that their error estimate was optimistic.

The turning point came with a space project known as GRACE. This joint mission between NASA and the German Space Agency, whose official name was Gravity Recovery and Climate Experiment, was launched in 2002 and ended in late 2017. GRACE consisted of a pair of satellites (dubbed Tom and Jerry) flying in close formation (200 kilometers apart) on polar orbits 500 kilometers above the Earth. Each satellite carried a satellite-to-satellite radar link to measure the distance between them very precisely, and a GPS receiver to track the orbit of each satellite separately. As the pair pass over a region with excess mass, such as a mountain, the gravitational attraction of the mass pulls the satellites toward each other, while if they pass over a region with a deficit of mass, such as a valley or depression, the satellites are pulled apart (see Figure 4.7). As the satellites passed repeatedly over different parts of the Earth they were able to map out its gravitational field in fine detail and with unprecedented accuracy. Furthermore, over the fifteen years of the mission, GRACE was able to measure time variations in the gravity field of the Earth. For example, it could measure the gravity variation caused

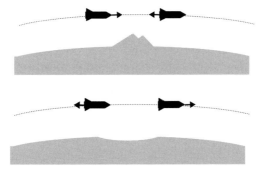

Fig 4.7 Measuring variations in Earth's gravity using GRACE. Top: As the two satellites pass over a mountain, its gravitational attraction pulls them toward each other. Bottom: As they pass over a large depression, the *absence* of mass causes them to separate a bit.

by seasonal changes in the amount of water in the Amazon and Ganges river basins, and the changes in gravity caused by mass loss in the ice sheets of Greenland and Antarctica. These measurements have obvious implications for water management and climate change. It could also monitor the rise of the North American and European land masses as they continue to "rebound" from the loss of the ice that had weighed them down during the last ice age.

With the dramatic improvements in accuracy for Earth's gravity field provided by GRACE, Ciufolini and his colleagues could now remove the Newtonian effects and uncover the tiny frame dragging effect more precisely. In 2010 and 2011, as GP-B was in the final phase of data analysis, Ciufolini and colleagues reported a result from LAGEOS in agreement with general relativity to about 10 percent, about a factor of two better than the final GP-B result.

Meanwhile, Ciufolini had succeeded in convincing the Italian Space Agency to go for a third laser-ranged satellite, called LARES (Laser Relativity Satellite), to be launched with an inclination of 69.5 degrees, very close to the required angle relative to LAGEOS I. However, the agency informed Ciufolini that, in order to reduce cost, it would not provide a launch vehicle powerful enough to achieve the same distance from the Earth as that of the two LAGEOS satellites, again preventing perfect cancelation of the Newtonian effect, which varies as the inverse cube of the radius of the orbit. Still, the advantage of having the proper inclination was enough to convince the LARES team to accept the launcher offered. LARES was launched on 13 February 2012, and in 2016, combining data from all three satellites with improved Earth data from GRACE, the LARES team reported a test of frame dragging at the 5 percent level.

The main difficulty in all of these precession experiments is that near Earth, gravity is just too weak! It would be wonderful to measure how gyroscopes precess near a black hole or a neutron star. The magnitude of the relativistic precession effects might be so large that one could see the precession "by eye." But this, of course, is completely impractical. The nearest black hole observed, V616 Monocerotis, is 3,000 light years away, and the nearest neutron star, Calvera, is between 250 and 1,000 light years away, so sending a gyroscope experiment there (maybe called "GP-X") is hopeless. But rotating neutron stars themselves can act as gyroscopes as effectively as can spheres of fused silica. And many neutron stars also act as "pulsars," emitting a beam of radio waves that we can detect, and some are in orbit around other stellar bodies. In fact, these celestial lighthouses have taken testing general relativity into a whole new realm, as we will see in the next chapter.

CHAPTER 5

Celestial Lighthouses for Testing Relativity

It is unlikely that Joe Taylor and Russell Hulse will ever forget the summer of 1974. It started uneventfully enough. Taylor, a young professor at the University of Massachusetts at Amherst, had arranged for his graduate student Hulse to spend the summer at the Arecibo Radio Telescope in Puerto Rico looking for pulsars. They had put together a sophisticated observational technique that would allow them to scan a large portion of the sky using the radio telescope in such a way that it would be especially sensitive to signals from pulsars. At that time around a hundred pulsars were known, so their main goal was to add new ones to that list, in the hope that, by sheer weight of numbers, they could learn more about this class of astronomical objects. But apart from the possible payoff at the end of the observations, the bulk of the summer would be spent in rather routine, repetitive observing runs and compilation of data that, as in many such astronomical search programs, would border on tedium. But on 2 July, good fortune struck.

On that day, almost by accident, Hulse discovered something that would catapult both Hulse and Taylor into the astronomical headlines, excite the astrophysics and relativity communities, and ultimately yield the first confirmation of one of the most important predictions of general relativity.

At least as far as relativists are concerned, the discovery ranks up there with the discovery of pulsars themselves. That discovery was equally serendipitous. In late 1967, radio astronomer Antony Hewish and his graduate student Jocelyn Bell at Cambridge University were attempting to study quasars by exploiting the phenomenon of scintillation, the rapid variation or "twinkling" of the radio signal that is caused by clouds of electrons in the solar wind out in interplanetary space. These variations are typically random in nature and are weaker at night when the telescope is directed away from the Sun, but in the middle of the night of 28 November 1967 Bell recorded a sequence of unusually strong, surprisingly regular pulses in the signal. After a month of further observation, she and Hewish established that the source of this signal was outside the solar system, and that the signal was a rapid set of pulses with a period of 1.3372795 seconds.

As a standard of time measurement, these pulses were as good as any atomic clock that existed at the time. It was so unexpected to have a naturally occurring astrophysical source with such a regular period that, for a while, they entertained the thought that the signals were a beacon from an extraterrestrial civilization. They even denoted their source LGM, for little green men. The Cambridge astronomers soon discovered three more of these sources, with periods ranging from a quarter to one and a quarter seconds, and other observatories followed with their own discoveries. The little green men theory was quickly dropped, because if the signal was truly from an alien civilization then it should have shown a Doppler shift as the alien planet orbited around its alien star. The only Doppler shift they saw was due to the Earth's motion around our Sun. The sources of these signals were renamed "pulsars" because of the pulsed radio emission.

This discovery had an enormous impact on the world of astronomy. The discovery paper for the first pulsar was published on 24 February 1968 in the British science journal *Nature*, and in the remaining ten months of that year over one hundred scientific papers were published reporting either observations of pulsars or theories of the pulsar phenomenon. In 1974, Hewish was rewarded for the discovery with the Nobel Prize in Physics, along with Martin Ryle, one of the pioneers of the British radio astronomy program. In some circles controversy still lingers over the decision of the Swedish Academy not to include Bell in the award. Now a renowned astronomer, academic leader and proponent of women in science, Dame Bell-Burnell ("Dame" being the female title that accompanies knighthood) has consistently expressed agreement with the Nobel decision.

Within a few years of the discovery, there was a general consensus about the overall nature of pulsars. Pulsars are simply cosmic lighthouses: rotating beacons of radio waves (and in some cases of optical light, X-rays and gamma rays) whose signals intersect our line of sight once every rotation period. The underlying object that is doing the rotating is a neutron star, a highly condensed body, typically a bit more massive than the Sun, but compressed into a sphere of around 20 kilometers in diameter, 500 times smaller than a white dwarf of a comparable mass, or 100,000 times smaller than a normal star of that mass. Its density is therefore about 500 million metric tons per cubic centimeter, comparable to the density inside the atomic nucleus. Neutron stars are so dense that a single teaspoon of neutron star matter on Earth would weigh the same as about 1,000 Great Pyramids of Giza in Egypt. As their name would indicate, neutron stars are made mostly of neutrons, with a contamination of protons and an equal number of electrons. Because a neutron star is so dense, it behaves as the ultimate flywheel, its rotation rate kept constant by the inability of frictional forces to overcome its enormous rotational inertia. Actually, there are some residual braking forces that do tend to slow it down, but an example of how small this effect can be is given by the original Bell–Hewish pulsar: its period of 1.3373 seconds is observed to increase by only 43 nanoseconds per year. Of the one hundred or so pulsars known by 1974, every one obeyed the general rule that it emits radio pulses of short period (between fractions of a second and a few seconds), and with a period that is extremely stable, except for a very, very slow increase. We will see that this rule almost proved to be the downfall of Hulse and Taylor.

Why a neutron star? Was this just a figment of the theorist's imagination, or was there some natural reason to believe in such a thing? In fact, neutron stars did begin as a figment of the imagination of the astronomers Walter Baade and Fritz Zwicky in the mid 1930s, as a possible state of matter one step in compression more extreme than the white dwarf state. This remarkable suggestion was made only a few years after the discovery of the neutron! Such highly compressed stars, they suggested, could be formed in the course of a supernova, a cataclysmic explosion of a star in its death throes, that occurs in galaxies throughout the universe, including our own. While the outer shell of such a star explodes, producing a flash of light that can momentarily exceed the light output of the entire galaxy and ejecting a fireball of hot gas, the interior of the star implodes until it has been squeezed to nuclear densities, whereupon the implosion is halted, leaving a neutron star as the cinder of the supernova. The neutron star should also be spinning very rapidly, for the following reason. All stars for which decent data exist are known to rotate, the Sun being the nearest example. Therefore, just as the figure skater spins more quickly when she pulls in her arms, exploiting the conservation of angular momentum, so too the collapsing, rotating core of the supernova should speed up.

Of the five supernovae in our galaxy of which we have historical records during the past thousand years, one occurred in the constellation Taurus in 1054. It was recorded by Chinese astronomers as a "guest star" that was so bright that it could be seen during the day. The remnant of that supernova is an expanding shell of hot gas known as the Crab Nebula. The observed velocity of expansion of the gas is such that, if traced backward in time for about 950 years, it would have originated in a single point in space. Several months after the discovery of the first pulsars, radio astronomers at the National Radio Astronomy Observatory trained the telescope on the central region of the Crab Nebula and detected radio pulses. The discovery was confirmed at the Arecibo observatory, and the pulse period was measured to be 0.033 seconds, the shortest period for a pulsar known at the time. Moreover, compared to other pulsars, the Crab pulsar was slowing down at an appreciable rate, around 10 microseconds in period per year. Put another way, the time required for the period to change by an amount comparable to the period itself is around 1,000 years, which is just the approximate age of the pulsar if it was formed in the 1054 supernova. Finally, if the pulsar is a rotating neutron star, the loss of rotational energy implied by its slowing spin turned out to be enough to keep the nebula of gas sufficiently hot to glow with the observed intensity.

The fact that all these observations were so consistent with one another provided a beautiful confirmation of the rotating neutron star model for pulsars. The most recent nearby supernova occurred in 1987 in the Large Magellanic Cloud, a small satellite galaxy to the Milky Way.

Other aspects of pulsars are not so clean cut or so simple, however, and one of these is the actual mechanism for the "lighthouse beacon," if indeed that is how the radio pulses are produced. In the conventional model, a pulsar is thought to have one important feature in common with Earth: its magnetic northern and southern poles do not point in the same direction as its rotation axis. On Earth, for instance, the geomagnetic northern pole is near Ellesmere Island, in the far north of Canada, not in the middle of the Arctic Ocean, as is the north pole of the rotation axis. There is one key difference, however. The magnetic field of a generic pulsar is a trillion times stronger than that of Earth. Such enormous magnetic fields produce forces that can strip electrons and ions from the surface of the neutron star and accelerate them to nearly the speed of light. This causes the particles to radiate copiously in radio waves and other parts of the electromagnetic spectrum, and because the magnetic field is strongest at the poles, the resulting radiation is beamed outward along the northern and southern magnetic poles. Because these poles are not aligned with the rotation axis, the two beams sweep the sky, and if one of them hits us, we record a pulse and call the source a pulsar. The precise details of this mechanism are still not fully worked out, partly because we have absolutely no laboratory experience with magnetic fields of such strength and with bulk matter at such monstrous densities. However, using massive computer simulations, researchers have recently been making progress in comprehending how the beams originate.

Nevertheless, by the summer of 1974 there was agreement on the broad features of pulsars. They were rapidly rotating neutron stars whose periods were very stable except for a very slow increase with time. It was also clear that the more pulsars we knew about and the more detailed observations we had, the better the chances of unraveling the details.

This is what motivated and guided Hulse and Taylor in their pulsar search. The receiver of the 1,000 foot radio telescope at the Arecibo observatory was driven so that as the Earth rotated, in one hour the instrument could observe a strip of sky a sixth of a degree wide by three degrees long. At the end of each day's observations the recorded data were fed into a computer, which looked for pulsed signals with a well-defined period. If a candidate set of pulses was found, it had to be distinguished from terrestrial sources of spurious pulsed radio signals, such as radar transmitters and automobile ignition systems. The way to do this was to return later to the portion of the sky to which the telescope was pointing when the candidate signals were received and see if pulses of almost exactly the same period were present. If so, they had a good pulsar candidate that they could then study further, such as by measuring its pulse period to the microsecond accuracy characteristic of other pulsars. If not, forget it and move on to another strip of the sky.

The day-to-day operation of the program was done by Hulse, while Taylor made periodic trips down from Amherst throughout the summer to see how things were going. On 2 July, Hulse was by himself when the instruments recorded a very weak pulsed signal. If the signal had been more than 4 percent weaker, it would have fallen below the automatic cutoff that had been built into the search routine and would not even have been recorded. Despite its weakness, it was interesting because it had a surprisingly short period, only 0.059 seconds. At the time, only the Crab pulsar had a shorter period. This made it worth a second look, but it was 25 August before Hulse got around to it.

The goal of the 25 August observing session was to try to refine the period of the pulses. If this were a pulsar, its period should be the same to at least six decimal places, or to better than a microsecond, over several days, because even if it were slowing down as quickly as the Crab was, the result would be a change only in the seventh decimal place. That is where the troubles began. Between the beginning and the end of the two-hour observing run, the computer analyzing the data produced two different periods for the pulses, differing by almost 30 microseconds. Two days later, Hulse tried again, with even worse results. As a result, he had to keep going back to the original discovery page in his lab notebook and cross out and re-enter new values for the period. His reaction was natural: annoyance. Because the signal was so weak, the pulses were not clean and sharp like those from other pulsars, and the computer must have had problems getting a fix on the pulses. Perhaps this source was not worth the hassle. If Hulse had actually adopted this attitude and dumped the candidate, he and Taylor would have been the astronomical goats of the decade. As it turned out, the suspicious Hulse decided to take an even closer look.

During the next several days, Hulse wrote a special computer program designed to get around any problems that the standard program might be having in resolving the pulses. But even with the new program, data taken on 1 and 2 September also showed a steady decrease of about 5 microseconds during the two-hour runs. This was much smaller than before, but still larger than it should be, and it was a decrease instead of the expected increase. To continue to blame this on the instruments or the computer was tempting, but not very satisfying.

But then Hulse spotted something. There was a pattern in the changes of the pulse period! The sequence of decreasing pulse periods on 2 September appeared to be almost a repetition of the sequence of 1 September, except it occurred 45 minutes earlier. Hulse was now convinced that the period change was real and not an artifact. But what was it? Had he discovered some new class of object: a manic depressive or bipolar pulsar with periodic highs and lows? Or was there a more natural explanation for this bizarre behavior?

The fact that the periods nearly repeated themselves gave Hulse a clue to an explanation. The source was indeed a well-adjusted pulsar, but it wasn't alone! The pulsar, Hulse postulated, was in orbit about a companion object, and the variation in the observed pulse period was simply a consequence of the Doppler shift (see Figure 5.1). When the pulsar is approaching us, the observed pulse period is decreased (the pulses are jammed together a bit), and when it is receding from us the pulse period is increased (the pulses are stretched out a bit). Actually, optical astronomers are very familiar with this phenomenon in ordinary stars. As many as half the stars in our galaxy are in binary systems (systems with two stars in orbit about each other), and because it

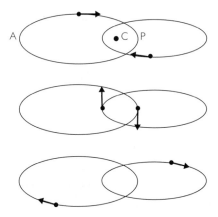

Fig 5.1 Orbit of a binary system such as the one containing the binary pulsar. The orbit of each body is an ellipse, and their velocities are shown here with arrows. The center of mass C of the system is the focus of each ellipse, while the periastron of one body is denoted P, and the apastron is A.

is rarely possible to resolve the two stars telescopically, they are identified by the up-and-down Doppler shifts in the frequencies of the spectral lines of the atoms in the atmospheres of the stars. In most ordinary stellar binary systems, the Doppler shifts of the spectra of both stars are observed; however, occasionally one of the stars is too faint to be seen, so astronomers can detect the motion of only one of the stars. And in recent years many exoplanets have been deduced by observing only the Doppler shifted spectra of the parent star. Such appeared to be the case here, where the pulse period was playing the same role as the spectral line in an ordinary star. One of Hulse's problems with this hypothesis was a practical one: he couldn't find any decent books on optical stellar binary systems in the Arecibo library because radio astronomers don't usually concern themselves with such things.

Now, because the dish of the Arecibo telescope is built into a natural bowl-shaped valley in the mountains of Puerto Rico, it could only look at the source when it was within 1 hour on either side of the zenith or overhead direction (hence the two-hour runs), and so Hulse couldn't just track the source for hours on end; he could only observe it during the same two-hour period each day. But the shifting of the sequence of periods in the 1 and 2 September data meant that the orbital period of the system must not be commensurate with 24 hours, and so each day he could examine a different part of the orbit, if indeed his postulate was right. On Thursday 12 September he began a series of observations that he hoped would unravel the mystery (see Figure 5.2).

On 12 September the pulse period stayed almost constant during the entire run. On 14 September, the period started from the previous value and decreased by 20 microseconds over the 2 hours. The next day, 15 September, the period started out a little lower and dropped 60 microseconds, and near the end of the run it was falling at the rate of 1 microsecond per minute. The speed of the pulsar along the line of sight must be varying, first slowly, then rapidly. The binary hypothesis was

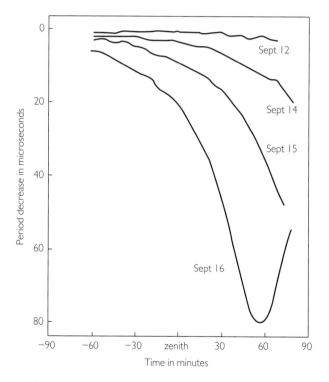

Fig 5.2 Pulse period changes of the binary pulsar over a five-day period in September 1974. Adapted from a page of Russell Hulse's notebook.

looking better and better, but Hulse wanted to wait for the smoking gun, the clinching piece of evidence. So far the periods had only decreased. But if the pulsar is in orbit, its motion must repeat itself, and therefore he would eventually be able to see a phase of the orbit when the pulse period increased, ultimately returning to its starting value to continue the cycle.

He didn't have long to wait. The very next day, 16 September, the period dropped rapidly by 70 microseconds, and with only about 25 minutes left in the observing run it suddenly stopped decreasing, and within 20 minutes it had climbed back up by 25 microseconds. This was all Hulse needed, and he called Taylor in Amherst to break the news. Taylor flew immediately down to Arecibo, and together they tried to complete the solution of this mystery. However, the real excitement was still to come.

The first thing they determined was the orbital period, by finding the shortest interval over which the pattern of pulse readings repeated. The answer was 7.75 hours, so the 45-minute daily shift that Hulse had seen was just the difference between three complete orbits and one Earth day.

The next obvious step was to track the pulse period variations throughout the orbit to try to determine the velocity of the pulsar as a function of time. This is a standard approach in the study of ordinary binary systems, and a great deal of information can be obtained from it. If we adopt Newtonian gravitation theory for a moment, then we know that the orbit of the pulsar about the center of mass of the binary system (a point somewhere between the two, depending upon their relative masses) is an ellipse with the center of mass at one focus (see Figure 5.1). The orbit of the companion is also an ellipse about this point, but because the companion is unseen, we don't need to consider its orbit directly. The orbit of the pulsar lies in a plane that can have any orientation in the sky. It could lie on the plane of the sky, in other words perpendicular to our line of sight, or we could be looking at the orbit edge on, or its orientation could be somewhere between these extremes. We can eliminate the first case, because if it were true, then the pulsar would never approach us or recede from us and we would not detect any Doppler shifts of its period. We can also forget the second case, because if it were true, then at some point the companion would pass in front of the pulsar (an eclipse) and we would lose its signal for a moment. No such loss of the signal was seen anywhere during the eight-hour orbit. So the orbit must be tilted at some angle relative to the plane of the sky.

That is not all that can be learned from the behavior of the pulsar period. Remember that the Doppler shift tells us only the component of the pulsar velocity along our line of sight; it is unaffected by the component of the velocity transverse to our line of sight. Suppose for the sake of argument that the orbit were a pure circle. Then the observed sequence of Doppler shifts would go something like this: starting when the pulsar is moving transverse to the line of sight, we see no shift; one-quarter period later it is moving away from us, and we see a negative shift in the period; one-quarter cycle after that it is again moving transverse, and we see no shift; one-quarter cycle later it is moving toward us with the same velocity, so there is an equal positive shift in the period; after a complete orbital period of seven and three-quarter hours, the pattern repeats itself. The pattern of Doppler shifts in this case is a nice symmetrical one, and totally unlike the actual pattern observed.

The observed pattern tells us that the orbit is actually highly elliptical or eccentric. In an elliptical orbit, the pulsar does not move on a fixed circle at a constant distance from the companion; instead it approaches the companion to a minimum separation at a point called periastron (the analogue of perihelion for the planets) and separates from the companion one-half of an orbit later to a maximum distance at a point called apastron. At periastron, the velocity of the pulsar increases to a maximum, and following periastron it decreases again, all over a short period of time, while at apastron, the velocity slowly decreases to a minimum value and afterward it slowly increases again.

The observed behavior of the Doppler shift with time indicated a large eccentricity (see Figure 5.3). Over a very short period of time (only two hours out of the eight) the Doppler shift went quickly from zero to a large value and back, while over the remaining six hours, it changed slowly from zero to a smaller value in the opposite sense and back. In fact, the 16 September smoking gun observation saw the pulsar pass through periastron, while the 12 September observations saw the pulsar moving slowly through apastron. Detailed study of this curve showed that the

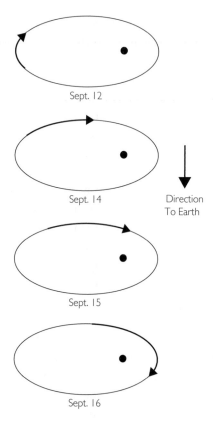

Fig 5.3 Location of the pulsar in its orbit. On 12 September, the pulsar is moving through apastron; its speed is low and slowly varying, so there is little change in the observed period (see Figure 5.2). The pulsar is moving away from us, so the period is longer than the "rest" period. On 14 September the pulsar is moving almost transversely, so there is little Doppler shift, and the period is shorter than before. On 15 September, the pulsar is starting to move toward us and is speeding up as it nears periastron; the pulse period decreases markedly toward the end of the run. On 16 September the pulsar starts out moving almost transversely, then quickly passes through periastron, so its velocity toward us quickly reaches a maximum, then decreases; the pulse period rapidly reaches a minimum and then increases. The portion of the orbit seen during the same two-hour interval each day varies because the orbital period is 7.75 hours, so the portion seen is 45 minutes further advanced each day.

separation between the two bodies at apastron was four times larger than their separation at periastron. It also showed that the direction of the periastron was almost perpendicular to the line of sight, because the periastron (the point of most rapid variation in velocity) coincided with the largest Doppler shift (the point where the pulsar has the smallest amount of transverse motion).

At this point, things began to get very interesting. The actual value of the velocity with which the pulsar was approaching us, as inferred from the decrease in its pulse period, was about 300 kilometers per second, or about one-thousandth of the speed of light! The velocity of recession was about 75 kilometers per second. These are high velocities! The speed of the Earth in its orbit about the Sun is only 30 kilometers per second. Combining the speed information with the orbital period, one could estimate that the average separation between the pulsar and its companion was only about as large as the radius of the Sun.

When news of this discovery began to spread in late September 1974 it caused a sensation, especially among general relativists. The reason is that the high orbital speeds and close proximity between the two bodies made this a system where effects of general relativity could be measurable.

In fact, even before Hulse and Taylor's discovery paper on the binary pulsar appeared in print (but too late to stop the presses), Taylor and his colleagues had detected one of the most iconic effects of general relativity, known as the "periastron advance" of the orbit.

According to Newton's gravitational theory, the orbit of a planet about its star is generally an ellipse, with the long axis of the ellipse always pointing in the same direction. For a binary star system, each body moves on an ellipse with the center of mass of the system as a focus (see Figure 5.1), but both ellipses are fixed in orientation. Any deviation from a pure Newtonian gravitational force between the two bodies, such as the gravitational tug of a third nearby body, or a modification of Newton's laws provided by a theory like general relativity, can cause the ellipse to rotate or "precess," as illustrated in Figure 5.4. As a result, the periastron, or point of closest approach, will not always be in a fixed direction, but will advance slowly with time.

In the case of Mercury orbiting the Sun, astronomers had already established by the middle of the nineteenth century that its point of closest approach, called the perihelion, was advancing at a rate of

575 arcseconds per century. It was reasonable to assume that this was the result of the perturbing effects of the other planets in the solar system (Jupiter, Venus, Earth, etc.), and French astronomer Urbain Jean Joseph Le Verrier, who was the director of the Observatory of Paris, set out to calculate these effects. In 1859 he announced that there was a problem. The sum total of the effects of the planetary perturbations fell short of the observed advance of the perihelion by about 43 arcseconds per century (here we are quoting the modern value of the discrepancy). Le Verrier had recently become famous by predicting that some anomalies in the orbit of Uranus were being caused by a more distant planet, a prediction confirmed months later when German astronomers, using his calculations as a guide, discovered Neptune. It was therefore natural for Le Verrier and his contemporaries to postulate that the anomaly

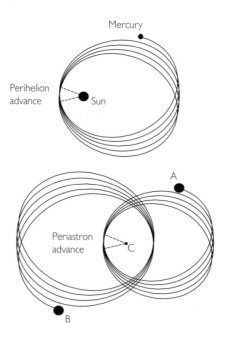

Fig 5.4 Top panel: Advance of the perihelion of Mercury's orbit around the Sun. Because the Sun is so massive, it barely moves. The rate of advance has been greatly exaggerated. Bottom panel: Advance of the periastron of two stars A and B of comparable mass orbiting around their center of mass C.

in Mercury's orbit was due to a planet between Mercury and the Sun. They even gave the planet the name Vulcan, after the Roman god of fire. Despite numerous claimed sightings of Vulcan over the next 60 years, no credible evidence for the planet was ever found.

Einstein was well aware of the problem of Mercury's anomalous perihelion advance, and in fact he used it as a way to test and ultimately reject earlier versions of his theory, notably a "draft" theory he had devised in 1913 with mathematician Marcel Grossmann. In November 1915, when everything seemed to be falling into place theoretically for his latest attempt, the tipping point occurred when his calculations showed that the theory gave the correct value for the missing perihelion advance. He later wrote that this discovery gave him "palpitations of the heart."

In 1915, the agreement that Einstein found was fairly crude, because the observations of Mercury's orbit were not very accurate. But since the 1970s Mercury's perihelion advance has become another high-precision confirmation of general relativity, made possible by high-precision radar tracking of planets and spacecraft. The most recent test was made using Mercury MESSENGER. In 2011, MESSENGER became the first spacecraft to orbit Mercury, and radar measurements of the orbiter were made until the spacecraft ended its mission in 2015 with a controlled crash on the surface of Mercury. The data from that mission yielded a measurement of Mercury's perihelion advance in agreement with general relativity to a few parts in 100,000. Improved measurements down to the level of parts per million may be possible with data from the joint European-Japanese BepiColombo mission to place two orbiters around Mercury, launched in late 2018.

If Einstein's theory indeed played a role in Hulse and Taylor's system, then measuring the binary analogue of Mercury's advance was of high priority, and during a two and a half month observing program that ended on 3 December 1974, Taylor and his colleagues tried to pin it down. Coming up was the seventh installment of the Texas Symposium on Relativistic Astrophysics that had begun in Dallas in 1963. After cycling twice through a trio of cities that included Austin and New York, it was back in Dallas. The data analysis was completed just in time for Taylor to reveal to the audience on 20 December that the rate of periastron advance for the binary pulsar was 4 degrees per year. This advance rate is about 36,000 times larger than the rate for Mercury, but that's not a surprise. Because of the higher orbital speed and smaller separation in the binary system, the raw effects of general relativity are roughly 100 times larger than for Mercury, and the binary system completes 250 times more orbits per year, so the cumulative effect on the periastron builds up faster. Taylor would return to the Texas symposium four years later with an even more impressive announcement.

While it was great to see Einstein's theory in action in a new and exotic arena outside the solar system, Taylor's measurement didn't actually provide a real test of the theory. The problem is that the prediction of general relativity for the periastron advance for a binary system depends on the total mass of the two bodies; the larger the mass, the larger the effect. It also depends on other variables, such as the orbital period and the ellipticity of the orbit, but these are known from the observations. Unfortunately, we do not know the masses of the two bodies with any degree of accuracy. All we know is that they are probably comparable to that of the Sun in order to produce the observed orbital velocity, but there is enough ambiguity, particularly in the tilt of the orbit with respect to the plane of the sky, to make it impossible to pin the masses down any better from the Doppler shift measurements alone. Well, if we can't test general relativity using the periastron advance measurement, what good is it?

It is actually of tremendous good, because we can turn the tables and use general relativity to weigh the system! If we assume that general relativity is correct, then the predicted periastron advance depends on only one unmeasured variable, the total mass of the two bodies. Therefore, the measured periastron advance tells us what the total mass must be in order for the predicted and measured values to agree. From the fall 1974 observations, the inferred total mass was about 2.6 solar masses. Eventually, the periastron advance could be measured so accurately, 4.226585 degrees per year, that the total mass of the system was pinned down to 2.8284 solar masses. This was a triumph for general relativity. Here, for the first time, the theory was used as an active tool in making an astrophysical measurement, in this case the determination of the total mass of a distant system to high precision.

The relativists' intuition that this system would be a new laboratory for Einstein's theory was confirmed. But there was more to come.

During the first few months of observations of the pulsar, it was realized that this was a very unusual pulsar, over and above it being in a binary system. Once the strange variations in its observed pulse period were seen to be due to Doppler shifts resulting from its orbital motion, these variations could be removed from the data, allowing the observers to examine the intrinsic pulsing of the object, as if it were at rest in space. Its intrinsic pulse period was 0.05903 seconds, but if it was slowing down, as do other pulsars, it was doing so at an unbelievably low rate. It took almost an entire year of observation to detect any change whatsoever in the pulse period, and when the data were finally good enough to measure a change, it turned out to be only a quarter of a nanosecond per year. This was 50,000 times smaller than the rate at which the Crab pulsar's period changed. Clearly, any friction that the spinning neutron star was experiencing was very, very small, a fact that was consistent with the observation that its radio signal is so feeble that Hulse almost missed it. At this rate, the pulsar would change its period by only half a percent in a million years. The steadiness and constancy of this pulsar made it one of the best timepieces the universe had ever seen! In later years, many rapidly spinning but very steady pulsars would be detected; we will return to them in Chapter 9.

This remarkable steadiness made it possible for the observers to change how they made the measurements. Instead of measuring pulse periods (the difference in time between adjacent pulses) and the changes induced in the periods by the Doppler shift and various relativistic effects, they were able to measure the arrival times of individual pulses. The pulsar was so steady that Taylor and his colleagues could keep track of the radio pulses as they came into the telescope, and even when they had to interrupt the observations for long periods of time, while they returned to their home universities for such "mundane" duties as teaching, or while the telescope was used for other observing programs, they could return to the telescope after such breaks and pick up the incoming train of pulses without losing track of a single beep. To see why this leads to a big improvement in accuracy, consider a simple example. Suppose you can measure arrival times to a basic accuracy of a hundredth of a second. If the pulse period is 1 second, that implies that you can measure the period to 1 percent. But now if you measure the arrival time of pulse number 1 to a hundredth of a second, and then wait 50 seconds and measure the arrival time of pulse number 51 to the same precision, you will have measured the combined period of 50 pulses to a hundredth of a second, or that of a single pulse to one part in 5,000. How do you know that it was pulse 51 and not pulse 39 or 78? Because knowing the period to one part in 100, you know that the error in arrival time after 50 seconds is less than a whole pulse period, and so the pulse that arrives can't be anything but number 51.

Eventually, this arrival-time technique allowed them to determine the characteristics of the pulsar and the orbit with accuracies that began to boggle the mind: for the intrinsic pulsar period, 0.059030003217813 seconds; for the rate at which the intrinsic pulse period was increasing, 0.272 nanoseconds per year; for the rate of periastron advance, 4.226585 degrees per year; for the orbital period, 27906.9795865 seconds. Because the pulsar period changes by the quoted amount in the last three digits each year, when scientists refer to a measured pulsar period they must also refer to a specific date when that value would be true; in this case, the conventional date is 11 December 2003.

There was more to this accuracy than just an impressive string of significant digits. It also yielded two further relativistic dividends. The first of these was another example of applied relativity, or relativity as the astrophysicist's friend. Beside the ordinary shift of the pulses' arrival times caused by the varying orbital position of the pulsar, there are two other phenomena that can affect it, both relativistic in nature. The first is the time dilation of special relativity: because the pulsar is moving around the companion with a high velocity, the pulse period measured by an observer foolish enough to sit on its surface (he would, of course, be crushed to nuclear density) is shorter than the period observed by us. In other words, from our point of view the pulsar clock slows down because of its velocity. Because the orbital velocity varies during the orbit, from a maximum at periastron to a minimum at apastron, the amount of slowing down will be variable, but will repeat itself each orbit. The second relativistic effect is the gravitational redshift, a consequence of the principle of equivalence, as we have already seen in Chapter 2. The pulsar moves in the gravitational field of its companion, while we the observers are at a very great distance; thus, the period of the pulsar is redshifted, or lengthened, just as the period (or the inverse of the frequency) of a spectral line from the Sun is lengthened. This lengthening of the period is also variable because the distance between the pulsar and the companion varies from periastron to apastron, and it also repeats itself each orbit.

The combined effect of these two phenomena is a periodic up and down variation in the observed arrival times, over and above that produced by the orbital position. But whereas the orbital motion changed the pulse period in the fifth decimal place, these effects, being relativistic, are much smaller, changing the pulse period only beginning at the eighth decimal place. It is extremely difficult to measure such a small periodic variation, given the inevitable noise and fluctuations in such sensitive data, but within four years of continual observation and improvement in the methods, the effect was found, and the size of the maximum variation was 184 nanoseconds in the pulse period. Again, as with the periastron, this observation does not test anything, because the predicted effect turns out to contain another unknown parameter, namely the relative masses of the two bodies in the system. The periastron advance gives us the total mass, but not the mass of each body. Therefore we can once again be "applied relativists" and use the measured value of this new effect to determine the relative masses. The result was 1.438 solar masses for the pulsar, and 1.390 solar masses for the companion, good to about 0.07 percent. The understanding and application of relativistic effects here played a central role in the first precise determination of the mass of a neutron star.

These results for the masses of the two bodies were also interesting because they were consistent with what astrophysicists thought about the companion to the pulsar. Because it has never been seen directly, either in optical, radio or X-ray emission, we must use some detective work to guess what the companion might be. It certainly cannot be an ordinary star like the Sun, because the orbital separation between the pulsar and the companion is only about a solar radius. If the companion were

Sun-like, the pulsar would be plowing its way through the companion's outer atmosphere of hot gas, and this would cause severe distortions in the radio pulses that must propagate out of this gas. Such distortions are not seen. Therefore, the companion must be much smaller, yet still have 1.4 times the mass of the Sun. Such astronomical objects are called "compact" objects, and astrophysicists know of only three kinds: white dwarfs, neutron stars and black holes.

The currently favored candidate for the companion is another neutron star, based on computer simulations of how this system might have formed from an earlier binary system of two massive stars that then undergo a series of supernova explosions to leave two neutron-star cinders. The fact that both masses turn out to be almost the same is consistent with the observation that in these computer models, the central core of the pre-supernova star tends to have a mass close to 1.4 solar masses. After the outer shell of each star is blown away, the leftover neutron stars each have about this mass. This mass is called the Chandrasekhar mass, after the astrophysicist Subrahmanyan Chandrasekhar, who determined in 1930 that this value was the maximum mass possible for a white dwarf (this discovery earned "Chandra" a share of the Nobel Prize in Physics in 1983). Because a pre-supernova core is similar in many respects to a white dwarf, it is not surprising that this special mass crops up here as well.

According to these models, the pulsar that Hulse detected was formed in the first supernova explosion, which left a spinning pulsar with a strong magnetic field but without really affecting the companion star. But the fate of such a pulsar in isolation is to spin down, causing its magnetic field to weaken to such a degree that it no longer generates a detectable pulsar beam. This pulsar followed this track. Meanwhile, the massive companion star evolved toward its own inevitable supernova explosion, but first it underwent an expansion of its gaseous atmosphere, a common occurrence in the evolution of massive stars. The pulsar skimmed across this atmosphere, getting spun up to a rapid rotational speed, just as a beach ball spins up as it skims across water. It ended up as a weakly magnetized, rapidly spinning pulsar with a weak pulsar beam, pretty much as Hulse detected. The companion star finally exploded, leaving a second pulsar, again without affecting the first pulsar, and eventually the companion pulsar spun down enough that its beam was too weak to detect. The final system, then, is an old neutron star "recycled" as a fast pulsar (the one Hulse detected), and a young neutron star but "dead" pulsar (the one Hulse did not detect).

But the biggest payoff of the binary pulsar was yet to come.

General relativity predicts that binary star systems emit gravitational radiation. We will devote most of Chapter 7 to a discussion of the history and nature of gravitational waves. For our purposes here, the main thing we need to know is that by 1974, gravitational radiation was an active subject, and relativists were dying to find some. Even though Joseph Weber of the University of Maryland had claimed detection of waves as early as 1968, later experiments by other workers had failed to confirm his results, and the general feeling was that gravitational waves had not yet been found. Therefore, when the binary pulsar was discovered, and it was seen to be a new laboratory for relativistic effects, it seemed like a godsend. The binary pulsar could be used in the search for gravitational waves.

But not in the obvious sense. Because the binary pulsar is 29,000 light years away, the gravitational radiation that it emits is so weak by the time it reaches the Earth, and is of such low frequency (about 6 cycles per day), that it is undetectable by any detectors of today or the foreseeable future. On the other hand, general relativity predicts that gravitational waves carry energy away from the system, and therefore the system must be losing energy. How will that loss manifest itself? The most important way is in the orbital motion of the two bodies, because after all, it is the orbital motion that is responsible for the emission of the waves. A loss of orbital energy produces a speed-up of the two bodies and a decrease in their orbital separation. This seemingly contradictory statement can be understood when you realize that the total orbital energy of a binary system has two parts: a kinetic energy associated with the motion of the bodies, and a gravitational potential energy associated with the gravitational force of attraction between them. So although a speed-up of the bodies causes their kinetic energy to increase, a decrease in separation causes their potential energy to decrease by about twice as much, so the net effect is a decrease in energy. The same phenomenon happens, for example, when an Earth satellite loses energy because of friction against the residual air in the upper atmosphere; as it falls toward Earth it goes faster and faster, yet its total energy is declining, being lost in this case to heat. In the case of the binary pulsar, the speeding up combined with the decreasing separation will cause the time required for a complete orbit, the orbital period, to decrease.

Here was a way to detect gravitational radiation, albeit somewhat indirectly, and a number of relativists pointed out this new possibility in the fall of 1974, soon after the discovery of the binary pulsar. As we will see in Chapter 7, the effects of gravitational radiation are exceedingly weak, and this was no exception. The predicted rate at which the 27,000 second orbital period should decrease was only on the order of some tens of microseconds per year. Although this was an exciting possibility, the small size of the effect was daunting, and some thought it would take ten to fifteen years of continual observation to detect it.

Now flash forward just four years, to December 1978: the Ninth Texas Symposium on Relativistic Astrophysics, this time in Munich, Germany (Munich is in the state of Bavaria, sometimes considered the Texas of Germany). Joe Taylor was scheduled to give a talk on the binary pulsar. Rumor had it that he had a big announcement, and only a few insiders knew what it was. Cliff knew because he was scheduled to follow Taylor to present the theoretical interpretation of his results. A press conference had been set up for later in the day. In a succinct, fifteen-minute talk (a longer, more detailed lecture was scheduled for the following day), Taylor presented the bottom line: after only four years of data taking and analysis they had succeeded in detecting a decrease in the orbital period of the binary system, and the amount agreed with the prediction of general relativity, within the observational errors. This beautiful confirmation of an important prediction of the theory was a fitting way to open 1979, the centenary year of Einstein's birth.

It turned out that the incredible stability of the pulsar clock, together with some elegant and sophisticated techniques for taking and analyzing the data from the Arecibo telescope that Taylor and his team had developed, resulted in such improvements in accuracy that they were able to beat by a wide margin the projected timetable of ten years to see the effect. These improvements at the same time allowed them to measure the effects of the gravitational redshift and time dilation, and thereby measure the mass of the pulsar and of the companion separately. This was important because the prediction that general relativity makes for the energy loss rate depends on these masses, as well as on other known parameters of the system, so they needed to be known before a definite prediction could be made. With the values of about 1.4 solar masses for both stars, general relativity makes a prediction of 75 microseconds per year for the orbital period decrease. The most recent analysis of the data shows agreement with the prediction to better than 0.2 percent. The 1993 Nobel Prize in Physics was awarded to Hulse and Taylor for the discovery of the system and for the confirmation of the existence of gravitational radiation.

These results triggered intensive searches for more binary pulsars at the world's largest radio telescopes, resulting in an explosion in the number of known pulsars. More than 2,600 pulsars are currently known. What was once "the" binary pulsar has now joined around 290 other pulsars in binary systems. Most are utterly uninteresting for general relativity because they are so widely separated that the effects of the theory are unimportant or undetectable. A few pulsars are known to have planets, presumably not particularly habitable. Only about a hundred of the binaries have orbital periods shorter than a day, which makes them potential fodder for relativity. Some have white dwarf companions, while others appear quite similar to the original Hulse–Taylor binary pulsar. Two systems stand out, however.

The first was discovered in 2003 by Marta Burgay and collaborators using the Parkes 64 meter radio telescope in Australia. This system turned out to be full of surprises. The pulsar's orbital period was a factor of three shorter than that of the Hulse–Taylor binary, making it even more compact and relativistically interesting. The measured periastron advance was a remarkable 17 degrees per year, indicating a total mass of about 2.6 solar masses. A few months after the discovery, followup observations detected weak pulses from the companion! This was the first (and so far the only) *double* pulsar detected. In fact the two pulsars fit the same theoretical profile as the Hulse–Taylor system: pulsar A was evidently an old recycled pulsar, spun up to a spin period of only 23 milliseconds, while pulsar B, with a longer spin period of 2.8 seconds and a very feeble pulsed signal, was an almost dead, younger pulsar.

Because the variations of the pulses of both pulsars could be tracked, the two orbits could be fixed with more certainty than was the case for the Hulse–Taylor system. From these observations, along with the periastron advance, it was possible to determine the two masses directly: 1.338 and 1.249 solar masses for the main pulsar and the companion pulsar respectively. This in turn could be shown to imply that the orbit was almost perfectly edge-on relative to the line of sight. As a consequence, once per orbit, the signal from the main pulsar would pass close to the companion neutron star, and would therefore experience the Shapiro time delay in its propagation. This delay was measured, and the results agreed completely with the predicted delay, based on the measured mass of the companion. The same effect presumably occurred for the signal from the companion passing by the primary, but the companion's pulsed signal was too weak and ragged to be useful for measuring such tiny effects. The effect of time dilation and the gravitational redshift on the primary's observed pulse period was measured, and agreed with the prediction of general relativity. The decreasing orbit period was also measured, and it too agreed with the theory's prediction for gravitational radiation energy loss, with a precision today even better than that for the Hulse–Taylor pulsar.

There was one more surprise. The radio pulses from the primary pulsar A were known to be partially eclipsed by the passage of the secondary pulsar B across the line of sight, and the on-off flickering of the signal during the eclipse occurred on a roughly 3 second timescale, the same as the rotation period of pulsar B. This was not because the signal hit the actual neutron star, but because it passed through the highly charged "magnetosphere" of the companion. This is a region of strong magnetic fields and charged particles that wraps around the magnetic equator of the neutron star, and is much larger in diameter than the star itself. It is shaped more like a donut or bagel, with the neutron star residing in the middle of the hole (see Figure 5.5). The pulsar beam from *B* propagates along the magnetic poles in a direction perpendicular to the plane of the donut. It is essentially the same as the magnetosphere of the Earth, except that the Earth's magnetosphere is strongly distorted by the solar wind. The dense cloud of charged particles that are trapped within the magnetosphere can absorb and scatter radio waves as effectively as the solid material of the neutron star itself. So, as pulsar *A* passes behind pulsar *B* during the orbital motion, its signal can be absorbed or not by the magnetosphere of pulsar *B*, depending on the orientation of the pulsar at that moment. This model beautifully explained the pattern of eclipses, except for one thing. Over many orbits of the system, the detailed shape of the eclipse pattern drifted with time, contrary to what the model predicted.

General relativity gave the solution. As we saw in Chapter 4, the spin of an object should change direction slowly with time as the object moves through the curved spacetime of another body. This effect, known as the "geodetic precession," was one of the two effects verified by the Gravity Probe-B experiment. If the spin direction of the pulsar changes slowly with time, then, as can be seen in Figure 5.5, the signals from pulsar *A* might encounter a fatter or thinner part of the magnetosphere as

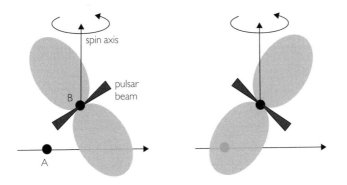

Fig 5.5 Pulsar *B* rotates about its spin axis with a period of 2.8 seconds, with its magnetic axis (which produces the pulsar beam) and donut-shaped magnetosphere (in gray) not aligned with the spin axis. Left: Pulsar *A* passes behind pulsar *B*, but the signal directed out of the page misses the magnetosphere. Right: After half a rotation of pulsar *B*, the magnetosphere now blocks the signal from pulsar *A*, causing an "eclipse."

time goes on. Incorporating this effect into the model gave complete agreement with the eclipse data, including the long-term changes, and the amount of geodetic precession of the spin needed to fit the data agreed with the prediction of general relativity. Sadly, general relativity also killed the "double" pulsar, because by 2008, the spin had precessed so much that the beam of pulsar *B* no longer passed across the Earth, and the system became a "single" binary pulsar. Pulsar *B* may reappear in 2035 when the wobbling spin brings its beam back into alignment with the Earth.

The second stand-out pulsar was discovered in 2014 by Scott Ransom and collaborators, using the Green Bank radio telescope in West Virginia. But instead of being in a binary system, this pulsar was in a *triple* star system, with two white dwarf companions. This is not as unlikely as it may seem. While the numbers are uncertain, as many as a fifth of the stars in the Milky Way could be in triple star systems. Alpha Centauri, our nearest neighbor, and Polaris, the North Star, are actually in triple systems, with two stars in a fairly close binary orbit and a more distant third star orbiting the pair. This new system is similar in that it consists of a close inner binary and a distant third star. The pulsar has a spin period of 2.73 milliseconds, and a mass of 1.44 solar masses. It is orbited once every 1.6 days by a white dwarf of only one fifth of a solar mass. This pair is orbited by a 0.4 solar mass white dwarf, with an orbital period of 327 days. Both orbits are almost perfect circles, and lie on the same plane (see Figure 5.6).

Everything about this system made it pretty useless for testing general relativity in the same manner as either the Hulse–Taylor binary or the double pulsar. The motions were too slow and the orbital separations too large for relativistic effects on the orbits to be very interesting. The orbits were too circular for the periastrons to be even located, let alone for their advances to be measured. The mass of the inner white dwarf was too small for the gravitational redshift of the pulsar signal to be measurable. The inclination of the orbits relative to the plane of the sky was found to be about 40 degrees, so the radio signal from the pulsar would pass nowhere near either white dwarf, making the Shapiro delay negligible. And finally, the emission of gravitational

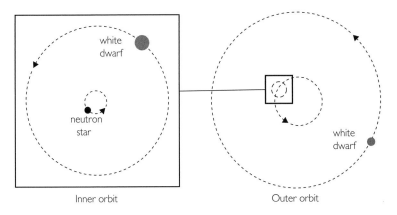

Inner orbit   Outer orbit

Fig 5.6 Pulsar in a triple system. Left: The neutron star is in a 1.6 day orbit with a low-mass white dwarf. Right: This inner binary is in a 327 day orbit with the other white dwarf.

radiation would be far too feeble for the decrease in the orbital periods to be detectable.

So why was this discovery so fantastic for general relativity? The answer goes back to the founding idea of general relativity, the principle of equivalence. In Chapter 2 we discussed how Einstein's insight about the equivalence between gravity and acceleration led him to a curved spacetime picture for gravitation. This was based upon the observational fact that bodies fall with the same acceleration, independently of their internal structure or composition. This idea is usually called the "weak equivalence principle," and was already appreciated in the ancient world. In 400 CE, Ioannes Philiponus wrote "... let fall from the same height two weights, of which one is many times as heavy as the other ... the difference in time is a very small one." Even before Galileo, this principle had been expounded and tested in the 1500s by Giambattista Benedetti and Simon Stevin, and if Galileo actually did drop things from the Leaning Tower of Pisa during his time there between 1589 and 1592, he was probably just demonstrating to his students what was by then a well-known concept. Even Isaac Newton carried out experiments using pendula to test this equality. As we noted in Chapter 2, the challenge of testing the weak equivalence principle to high precision was taken up by

Eötvös at the turn of the twentieth century. Because this principle is so crucial for the foundation of Einstein's theory, the effort to test it to higher and higher precision has continued to the present day. The current state of the art comes from two sources. One is a series of experiments carried out by a group headed by Eric Adelberger at the University of Washington in Seattle, called the "Eöt-Wash" experiments, that show that different materials fall with the same acceleration to a few parts in ten trillion. The other is a space experiment called MICROSCOPE, launched in 2016 by the French Space Agency, that yielded a limit of parts in a hundred trillion.

One interesting and important conclusion can be drawn from these results. Recall that the mass of an atomic nucleus is made up of the masses of the individual neutrons and protons, but that's not all. These neutrons and protons are held together by the strong forces that bind the nucleus. Einstein has taught us through special relativity that energy and mass are different manifestations of the same thing. Therefore, the mass of the atomic nucleus is made up of the sum of the masses of the individual neutrons and protons, and the "mass" associated with the binding energy. Now, since different elements have different amounts of nuclear internal energy per unit mass, and since experiments tell us that the different kinds of nuclei fall with the same acceleration, then the energy of the nuclear forces must "fall" with the same acceleration as do the nuclear particles themselves. A similar conclusion applies to the electromagnetic energy associated with the forces between the charged protons and electrons. So it would seem that not only do the fundamental building blocks of matter, such as protons, neutrons and electrons, fall with the same acceleration, but so do the various forms of energy associated with their interactions with each other, such as nuclear, electromagnetic and weak interactions.

But the standard model of fundamental particles tells us that there is a fourth interaction: gravitation. What about the energy associated with it? Does gravitational energy fall with the same acceleration as matter and the other forms of energy? The experiments we have just described don't provide an answer, because the internal gravitational energy of the laboratory-scale bodies employed in those experiments is utterly negligible. To get a meaningful amount of gravity, you need a large amount of mass, and therefore to test the equivalence principle including gravitational energy you need objects like planets or stars.

The first person to contemplate this possibility was Kenneth Nordtvedt. Born in Chicago, he received an undergraduate degree from MIT, took a Ph.D. degree at Stanford University, and had post-doctoral research positions back in the Boston area at Harvard and at MIT. But by 1965 he had developed a dislike for the lifestyle and politics of big cities, especially on either of the coasts, and had resolved to head for the heartland of America. When offered an assistant professorship at the then small Montana State University in tranquil and beautiful Bozeman, he accepted readily and headed west to begin his academic career in earnest.

Although his Ph.D. thesis was in solid-state physics, around 1967 he turned his attention to gravity and asked whether a massive body with its own internal gravity, such as the Earth, would fall in an external gravitational field with the same acceleration as, say, a ball of lead. To try to answer this question, Nordtvedt devised a way of treating the motion of planetary-size bodies that would be valid in any curved spacetime theory of gravity, or at least in a broad class of such theories. The equations he developed could encompass general relativity, the Brans-Dicke theory, then the leading alternative theory to Einstein's, and many others, in one fell swoop. To find the actual prediction of a chosen theory, such as general relativity, all one had to do was to specialize the equations by fixing the numerical values of certain coefficients that appeared in them. The calculations were complicated, with many, many terms in the final equation describing the acceleration of a massive body, but when all was said and done, two remarkable results emerged.

First, when the equations were specialized to general relativity there was a tremendous cancelation of terms, and the result was that different massive bodies would have *exactly* the same acceleration, regardless of how much internal gravity they possessed. Therefore, in general relativity the acceleration of gravitationally bound bodies was predicted to be the same as that of laboratory-size bodies. This beautiful prediction of general relativity, the equivalence of acceleration of bodies from the smallest to the largest sizes, is sometimes called the strong equivalence principle. Later research would show that this equivalence also applies to neutron stars and even black holes.

There was another remarkable result of Nordtvedt's calculations. In most other theories of gravity, including that of Brans and Dicke, the complete cancelation did not occur and a small difference in acceleration remained, depending on how strongly bound by internal gravity the bodies were. Therefore, even though these theories guaranteed that laboratory-size bodies fall with the same acceleration, satisfying the weak equivalence principle, as soon as one considered bodies with significant amounts of self-gravitational binding, the bodies would fall differently. In other words, in such theories, gravitational energy falls at a slightly different rate than mass and other forms of energy, such as nuclear energy, electromagnetic energy and so on. Thus, theories such as the Brans–Dicke theory were compatible with the weak equivalence principle, but *not* compatible with the strong equivalence principle. Today this is called the Nordtvedt effect.

Nordtvedt then proposed to search for this effect in the motion of the Moon. Consider the acceleration of the Earth and the Moon in the field of the Sun (see Figure 5.7). The gravitational energy per unit mass of the Moon is about one twenty-fifth that of the Earth, so they could in principle fall with different accelerations because the Earth is more tightly bound by its own gravity than is the Moon. Suppose, for the sake of argument, that the Moon falls with a slightly larger acceleration than the Earth (whether it is larger or smaller depends on the theory of gravity). The Moon orbits the Earth, but is being accelerated toward the Sun slightly more strongly than the Earth is; therefore, on each succeeding orbit the Moon is pulled a little closer to the Sun. What started out as a nearly circular orbit becomes elliptical, and on each orbit the ellipse becomes more and more elongated toward the Sun, until the Moon is pulled catastrophically from the hold of the Earth and plunges with a great splash into the Sun. Is the Nordtvedt effect a lunar calamity? Actually not, because we have forgotten an important fact: the Sun is in orbit about the Earth (as seen from the Earth's frame, of course). Thus, just as the Moon's orbit is elongated toward the Sun on one revolution of the Moon, by the next lunar revolution, 27 days later, the Sun has

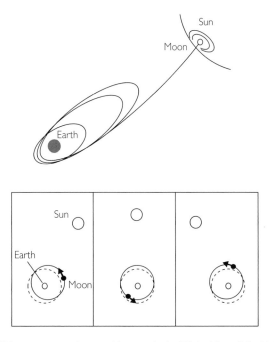

Fig 5.7 Lunar catastrophe or orbit perturbation? If the Moon fell with larger acceleration than the Earth toward the Sun, its orbit would become progressively more elongated until it was pulled into the Sun. But because the Earth–Sun orientation is changing because of the Earth's orbital motion, the elongation never builds up and instead merely produces a shifted orbit that always points toward the Sun (solid curves). In general relativity, the elongation does not occur at all (dashed curves).

moved by about 27 degrees in its orbit (the Sun's rate of revolution about the Earth is 360 degrees in 365 days, or about 1 degree per day), so on the next revolution, the elongation must occur in a direction toward the new position of the Sun. On the following revolution of the Moon, the elongation must be directed toward a still newer position, and so on. Therefore, instead of building up to a disastrous size, the elongation of the Moon's orbit that would be caused by the Nordtvedt effect maintains a fixed size, but is always oriented with its long axis toward the Sun. If the Moon were predicted to fall with a slightly smaller acceleration than the Earth, then the elongation would be in the opposite direction, with its long axis directed away from the Sun. If, as in general relativity, the two fall with the same acceleration, there would be no predicted elongation of this sort.

The crucial question is how large this effect might be. When Nordtvedt put in all the numbers, he found that the size of the elongation in the Brans–Dicke theory, for instance, could be as large as 1.3 meters, or about 4 feet. In general relativity, of course, the effect was zero. While this may seem like a ridiculously small effect, it would soon become eminently measurable.

After Apollo 11 astronaut Neil Armstrong took his first step on the Moon on 21 July 1969, he had a number of tasks to perform, one of which was to walk a few hundred meters from the lander and place on the lunar surface a device called a "retroreflector," an early version of the retroreflectors used on the LAGEOS satellites (page 101). This was a flat surface embedded with cube corner reflectors that could take a laser beam sent from Earth and reflect it back in the same direction from which it came. One could then send a short laser pulse from Earth, have it bounce off the retroreflector and return to Earth. Measuring the round-trip travel time of the pulse would give a measure of the Earth–Moon distance, with a precision that was expected to be of the order of 100 centimeters, easily sufficient to look for a possible Nordtvedt effect. Within a week and a half of deployment of the Apollo 11 retroreflector, astronomers at Lick Observatory in California had succeeded in bouncing laser pulses off it, and measuring the round-trip travel time to a precision corresponding to several meters. Subsequently, four other retroreflectors were placed on the Moon, two US devices, deployed by astronauts during Apollo 14 and 15, and two French-built reflectors, deposited during the Soviet unmanned missions Luna 17 and 21. By 1975, analyses of the laser ranging data showed absolutely *no* evidence of the Nordtvedt effect, to a precision of 30 centimeters, to the delight of Nordtvedt and anybody who feels that general relativity is correct. As Nordtvedt was fond of saying, "scientifically, zero can be just as important a number as any other."

Today, lunar laser ranging is carried out at over forty observatories worldwide, with ranging precisions in the millimeter regime, yielding important science about the Earth–Moon orbit, the rotation of the

Moon, continental drift on Earth, and even whether Newton's constant of gravitation is constant in time (it is, to the uncertainty in the measurement, which is a few parts in ten trillion per year). And recent analyses have continued to show no evidence of the Nordtvedt effect. One way to summarize the results is to state that the Earth and Moon fall toward the Sun with the same acceleration to a few parts in ten trillion, comparable to the limits achieved by tests of the weak equivalence principle, such as the Eöt-Wash and MICROSOPE measurements.

The pulsar in a triple system carried this test of the Nordtvedt effect into a new realm of extreme gravity. The system is a variant of the Earth–Moon–Sun system, with the inner neutron star / white dwarf binary substituting for the Earth and Moon, and the outer white dwarf substituting for the Sun. The mass relationships are different, with the Sun dominating the masses in the solar system case, and the neutron star dominating in the pulsar case. But the question is the same: do the neutron star and its white dwarf companion fall with the same acceleration toward the distant white dwarf?

The crucial difference is this. Whereas the internal gravitational binding energy of the Earth and the Moon represent only about a billionth of their total mass, the gravitational energy of a neutron star represents as much as 15 percent of its total mass. In other words, if you could somehow go into the neutron star and count up all the protons, neutrons, electrons and any other exotic particles that you might find, multiply each by its mass and add it all up, you would get something like 1.6 solar masses. The actual measured mass is 1.4 solar masses. The difference of about 0.2 solar masses, multiplied by the square of the speed of light, is the gravitational binding energy, and is in fact a negative number.[1] This is analogous to the phenomenon by which the mass of a helium

---

[1] If you are uncomfortable with the notion of negative energy, think of it in terms of bookkeeping. If you have a total amount of money, but you also owe something (say an automobile loan), that debt counts as a negative, reducing your total assets. But if you get an infusion of money (say by working overtime) and pay off the loan, you can own the car free and clear. In the same way, if you inject a lot of energy into the neutron star, enough to "pay off" the negative binding energy, you can free all the protons and neutrons. For all bound systems, from nuclei to neutron stars, physicists think of binding energy as negative because it simplifies the mass–energy bookkeeping.

nucleus is slightly smaller than the mass of four hydrogen nuclei (protons), so that when four protons fuse to form helium in the Sun, that mass difference becomes the energy on which we rely (in this example, the binding energy comes from the strong nuclear interactions). The binding energy of a typical white dwarf is parts in ten thousand of its mass, much larger than that of the Earth or Moon, but much smaller than that of a neutron star.

Soon after the pulsar in a triple was discovered, the team set about looking for the Nordtvedt effect. If, for example, the neutron star were to fall with larger acceleration than its companion, then its orbit would be displaced slightly toward the distant white dwarf, while the orbit of the inner white dwarf would be displaced slightly in the opposite direction, and the displacement would rotate in time to track the distant companion (as in Figure 5.7). In July 2018, Anne Archibald, who led the data analysis, announced that they had found no evidence in the pulsar signal for such orbital dispacement of the pulsar. The data showed that the neutron star and the white dwarf fall with the same acceleration to about three parts in a million, showing no Nordtvedt effect, and in complete agreement with general relativity.

Because the strong internal gravity of the neutron star *could have* produced an anomalous effect (and does, in many alternative theories), this constitutes a beautiful test of general relativity in the strong-field regime. Even more extreme tests of Einstein's theory are possible, but for these, we must first discuss black holes, as we do in the next Chapter.

CHAPTER 6

How to Use a Black Hole to Test General Relativity

The black hole is probably the most bizarre, exotic and fascinating prediction of Einstein's general relativity. This object, composed purely of warped spacetime, endowed with the ability to trap anything, from light rays to Marvel superheroes, that crosses its famous "event horizon," has lodged itself in the public imagination like no other aspect of physics. Ask the average person on the street to name the four fundamental forces of nature or the basic particles that make up the atom, and more often than not you will get a blank stare. But ask that person to say something about black holes, and you may well get a fairly coherent summary of their basic properties. You will almost certainly hear the name Stephen Hawking, the person universally attached to the black hole idea, and you may also get a list of movies that have featured black holes.

But even more bizarre is that we know, almost as certainly as one knows anything in science, that black holes exist. In Chapters 7 and 8 we will learn about the spectacular evidence for merging black holes gleaned from the gravitational waves they emitted. But long before those detections, the evidence for black holes obtained by astronomers using light, ranging from gamma rays to radio waves, was so solid that there was very little doubt that these things are really out there. Our goal here is to describe black holes and the evidence (apart from gravitational waves) for them. More importantly, we will explain to you how we may be able to carry out some remarkable new tests of general relativity using black holes.

We have already seen in Chapter 3 how John Michell and Pierre Simon Laplace in the late 1700s speculated on the possibility of a body so dense that it could prevent light from escaping its surface and reaching great distances. While it is tempting to say that they "predicted" black holes, the fact is that the physics they used—Newtonian gravity and the corpuscular theory of light—was not correct in the end. We now know that gravity is governed by general relativity (at least all the evidence supports that so far), and that light is governed by Maxwell's equations, even if you choose to describe it using the quantum concept of the photon. Nevertheless, it is amusing to learn about the ideas that these great Enlightenment thinkers came up with using their imaginations and the accepted physics of their day.

The *modern*, relativistic story of black holes began on the battlefields of World War I, with a soldier-scientist named Karl Schwarzschild. He was born in Frankfurt, Germany, on 9 October 1873, the son of a banker. The family traced its roots to medieval Frankfurt, when Jews were confined to a ghetto, but under the protection of the king, or "Kaiser." Even then, the Schwarzschild family was well-to-do. The family name evidently originates in a tradition of that period of identifying a family by a plaque ("schild" in German) that it would affix to the front of its house. For Karl's ancestral family the plaque was black ("schwarz"), hence the name. One of their neighbors in the ghetto had a plaque that was red ("rot"). The descendants of that family would become the famous Rothschild family of financiers. By 1811 the Frankfurt ghetto was abolished and some measure of civil rights was granted to Jews.

Schwarzschild showed an early interest in science, and by the age of 23 had received a Ph.D. from the Ludwig Maximillian University in Munich. His career advanced rapidly, with posts in Vienna and Göttingen, culminating in 1909 in the directorship of the Astrophysical Observatory in Postdam, one of the premier observatories of Europe at the time. He contributed seminal work to a wide range of topics, including solar physics, the statistics of stellar motions, the optics of astronomical lenses, the determination of the orbits of comets and asteroids, and the classification of stars by their spectra. He even traveled to Algiers to study the solar eclipse of 1905. This was long before the bending of light was an issue, of course; in those days eclipses were used mainly to study the corona of the Sun, and to search for the hypothetical planet Vulcan between the Sun and Mercury.

When World War I broke out in August 1914 he was 41 years old, and therefore not subject to the draft, nor required to enlist. But, like many Jews in Germany at the time he volunteered, desiring to demonstrate the loyalty of the Jewish community to Germany (other Jewish scientists who served this way included Fritz Haber, James Franck and Gustav Hertz). He never fought on the front lines, but because of his technical background he was assigned to head a meteorological station at Namur, Belgium, and then was attached to the artillery staff in France and later on the Eastern front. He even wrote a paper in 1915 on the effects of air resistance on the path of a projectile, but publication was held up for security reasons.

But in late 1915, at the Russian front, he contracted a rare and painful autoimmune skin disease called pemphigus, that even today is difficult to treat. While undergoing treatment he received copies of the papers that Einstein had delivered a few weeks earlier at the Prussian Academy of Sciences, describing his new theory of general relativity. Schwarzschild recognized that the theory was very complicated mathematically, but he set out to see if, by making some simplifying assumptions, he could find a solution to the Einstein equations of general relativity.

For his first attempt, he assumed that the sought-after solution would be static, or unchanging in time. He then assumed that the solution would be spherically symmetric, or the same no matter how you rotated the system about a central point. He imagined that the solution might apply to a perfectly spherical body, such as a star, sitting at rest, and isolated or far away from any other perturbing bodies. For his first attempt at a solution, he further assumed that the body was so small in size that you could ignore its extent or its internal structure. He called it a "Massenpunkt," or point mass, an object whose mass was entirely concentrated at a point. This kind of assumption is routinely used in physics because it allows you to get an idea of what the solution outside the body looks like, without having to worry about the messy details of the body's internal structure. To his surprise, Schwarzschild found a simple exact solution to Einstein's complex equations.

For his second attempt, he assumed that the body had a finite size, but that its interior density (the amount of mass per unit volume) was constant throughout the body. This is not a totally realistic assumption, because we know that bodies bound by their own gravity, such as the Earth or the Sun, are more dense at the center than at the surface. But once again, such an assumption might give useful insights without having to sweat a lot of messy details such as whether the body is solid, liquid or gas (or a combination), or whether it is hot or cold. Here again, he found an exact solution to the Einstein equations.

He wrote up his solutions in two papers and sent them to Einstein, who communicated the Massenpunkt paper to the Prussian Academy of Sciences on 13 January 1916, and the finite-body paper on 24 February. Einstein was amazed that Schwarzschild had managed to obtain these exact solutions, and pronounced himself very pleased with the second paper. The Massenpunkt paper, not so much.

The Massenpunkt paper had a feature that troubled Einstein greatly. Schwarzschild's solutions gave formulae for the warped geometry of spacetime that varied as a function of the distance from the center. Very far from the body, the geometry became that of ordinary flat space and time, as you would expect when the gravity of the body becomes negligible at great distances. As you approach the body, the geometry becomes more and more warped, again as expected when gravity gets stronger. But in the Massenpunkt solution, when the distance from the center reaches a value given by twice the mass of the point (multiplied by Newton's constant of gravitation and divided by the square of the speed of light), things got crazy. One function became infinite (one divided by zero), while another became zero. This behavior is what physicists call a "singularity," and for years afterward this was called the "Schwarzschild singularity." Since it occurs wherever the distance from the center has this special value, it is actually a spherical surface surrounding the point mass. It's not a surface made of matter that you could bounce off or crash into, but is rather an imaginary surface or boundary that we define mathematically because odd mathematical things happen to the solution on that surface. The question was, what?

Einstein believed that such singularities were unacceptable (we will learn in Chapter 7 that he thought he had disproved gravitational waves in 1938 because of the apparent existence of singularities in the solution), and that such a Massenpunkt would therefore not occur in nature. On the other hand, Schwarzschild's second solution for an extended body was completely acceptable. The geometry of spacetime became more warped as one approached the body (in fact, for the same mass, the external solution was identical to the solution for the point mass), but once inside the body, the solution changed and was perfectly finite all the way to the center. Einstein and others who examined Schwarzschild's two solutions believed that there must be something in the laws of physics to prevent a body from ever being so small in relation to its mass that it would reside completely inside this special "Schwarzschild" radius. To get a sense of scale, for an object with the mass of the Earth, the Schwarzschild radius is about a centimeter, roughly the size of the tip of your baby finger. For an object with the mass of the Sun, it is about three kilometers. For what we today would call a fifty million solar mass black hole, it is the radius of the Earth's orbit around the Sun. A body would have to be compressed to an incredibly small size and an enormous density to fit inside its Schwarzschild radius. To Einstein and his contemporaries, the world seemed safe from Schwarzschild's horrible singularity.

Unfortunately, nothing could save Schwarzschild from the ravages of pemphigus; he died on 16 May 1916. Even though he was German, he was such a famous astronomer that *The Observatory* in England published an obituary that August. It described his many contributions to astronomy and astrophysics, but made no mention of his solution to the Einstein equations of general relativity. Einstein's theory was still too new and obscure.

And that pretty much was it for black hole research for the next forty years. A few papers related to Schwarzschild's solution were published here and there, but they were largely ignored. Ironically, a young student named Johannes Droste, working under the tutelage of the great Dutch physicist Hendrik Lorentz, found the Massenpunkt solution in the spring of 1916, totally independently of Schwarzschild. But his paper was published in the relatively obscure *Proceedings of the Royal Netherlands Academy of Sciences* (in Dutch), and wasn't "discovered" until many decades later. During the 1920s and 1930s, a few researchers, including Eddington and Howard P. Robertson, tried to explore what was really going on at the Schwarzschild singularity. Was this a place where bizarre physics occurred, or was the singular behavior merely an artifact of the coordinates that Schwarzschild used, in the same way that the coordinates of latitude and longitude on Earth are singular at the north and south poles where all the lines of longitude meet at one point. Some of these papers, seen in retrospect, held important clues, but they were barely noticed.

In 1939, the American theoretical physicist J. Robert Oppenheimer and his student Hartland Snyder published a remarkable paper entitled "On continued gravitational contraction." In it they showed that a massive enough star that runs out of thermonuclear sources of energy will no longer be able to support itself against the crushing force of its own gravity, and will contract. Using Einstein's equations, they showed that the decreasing radius of the star would reach the Schwarzschild radius, but the star would continue to shrink. An observer riding on the surface of the star as it collapses inward would not observe any "singular" or bizarre behavior during the contraction, while observers at great distances would observe light emitted from the star's surface becoming progressively redder (the gravitational redshift effect) and fainter until, after a long time, the star's light would be essentially undetectable. For all practical purposes it would be "black," and the final object would be described by Schwarzschild's Massenpunkt solution. Eighty years later, this paper reads like a modern paper on black hole physics, with many of the insights into the nature of these objects that we have come to understand. But at the time, it also had almost no impact. Oppenheimer never followed up on the paper, and within three years would turn all of his attention to leading the Manhattan Project to develop an atomic bomb.

We must remember that this was a period when general relativity was considered a backwater of physics. Very few people worked in the field.

The noted general relativity theorist Peter Bergmann, who had been an assistant of Einstein during the late 1930s, once joked that if he ever needed to find out *everything* that was going on in general relativity, he only needed to call up his five best friends. Physicists were much more concerned with quantum mechanics, atomic and nuclear physics, field theory and elementary particles, and, after World War II, with developing the new technologies that had emerged from wartime research, such as radar, nuclear power, transistors, semiconductors, masers and lasers.

In 1956, Martin Kruskal, a mathematical physicist working in the plasma physics laboratory at Princeton University, realized that he could find a new system of coordinates in which Schwarzschild's "singularity" would disappear. He described his discovery to his Princeton colleague John Wheeler, who had begun to take an interest in general relativity. Wheeler thought it was nice, but otherwise paid little attention to it. But by 1959, Wheeler suddenly realized the significance of the discovery and wrote a paper with Kruskal's name as the author and submitted it to *Physical Review*. He somehow neglected to tell Kruskal what he was up to. A few months later, while on sabbatical in Germany, Kruskal received out of the blue the galley proofs of a paper that he didn't know he had written. But he recognized the figures in the paper as being in a style typical of his friend Wheeler, and he urged him to be a co-author. Wheeler declined, and Kruskal's paper became one of the foundations of a new understanding of the nature of black holes.

At the same time, George Szekeres, a Hungarian mathematician working at the University of Adelaide, Australia, and David Finkelstein, an American physicist at the Stevens Institute of Technology in New Jersey, were also working on similar approaches to resolving Schwarzschild's singularity.

These researchers showed conclusively that nothing "singular" or infinite happens at the Schwarzschild radius. A light ray, an atom or a graduate student can head toward the object and cross the Schwarzschild radius without experiencing anything infinite. To be sure, there will be the inevitable stretching and squeezing of a body as it approaches the object. This is nothing more than the same kind of tidal effects that the Moon induces upon the Earth, for example, stretching it along the line directed toward the Moon, and squeezing it in the perpendicular directions. These forces can be large enough to squeeze and stretch the poor graduate student into a long thin noodle, but no matter what, the forces remain *finite* as the student crosses the "magic" radius.

The true significance of the Schwarzschild radius turned out to be rather different, and quite astonishing. The sphere defined by this special radius turns out to be the boundary between two realms. Outside the sphere is the normal external universe, where people can travel freely, subject to the speed of light limitation, and can communicate with each other using light signals. You can even safely go into orbit around the object.

Once you cross the Schwarzschild radius, however, your fate is very different. Escape is impossible. You can fire up the most powerful rocket imaginable, subject only to the normal laws of physics, but you will be unable to get out. You are pulled inexorably toward a point at the center of the object, there to be squashed to zero volume and infinite density. In desperation you send a light signal outwards, pleading for help, and indeed you witness the signal moving away from you at the speed of light. But, unbeknownst to you, that signal is actually following you inward, later to join you and everything else that ever crossed that fatal sphere in a crushing finality.

It may seem contradictory to imagine sending a light signal outward, yet to have that signal actually follow you inward. One analogy that explains how this might work is to imagine a swimmer who always swims at a fixed speed within water, never faster, never slower. The swimmer is in the Niagara river, just above the famous waterfalls (Figure 6.1). Because her speed is higher than that of the current, she can freely swim up the river, down the river or across the river. Compared to a person treading water and following the current, she is always swimming at the same speed. But if she allows herself to go over the falls, her fate is different. She can try to swim upwards, and indeed relative to the person floating freely with the descending water, she is moving upward at her normal speed. Yet both swimmers are moving downward, to be dashed on the rocks below. Like all the analogies used in this book, this one is

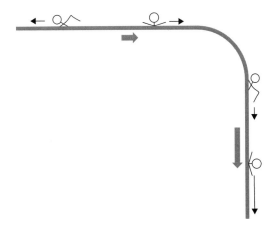

Fig 6.1 A waterfall as an event horizon. Above the waterfall a person floats with the current, while a swimmer swims away from him at her fixed speed, and slowly makes her way upstream. Below the waterfall the person still flows with the current, and the swimmer still moves away from him at her standard speed, but now both are falling to the rocks below, and the swimmer will never reach the top.

not perfect, but it gives a sense of how light can never escape from the Schwarzschild object. Yet contrary to what John Michell thought (see page 29), light never actually comes to rest according to anybody who measures its speed.

Because the Schwarzschild radius is the boundary between what can and cannot communicate with the outside world, it came to be called the "event horizon." Just as you cannot receive light from the Sun after it falls below the Earth's horizon, you cannot receive any signal from any event that occurs inside the Schwarzschild radius.

By the early to mid 1960s, these kinds of results convinced many general relativists that Schwarzschild's Massenpunkt solution was something to take seriously. John Wheeler was one of them, and in fact the term "black hole" is often credited to him. He had been ruminating on an appropriate term for these objects, and during a 1967 lecture he was giving at the Goddard Center for Space Studies in New York, he wondered aloud about a suitable name. Somebody in the audience shouted "black hole," and Wheeler immediately adopted and promoted it.

But to most physicists and almost all astronomers, black holes were curiosities of Einstein's theory, but so what? That attitude began to change with the discovery of quasars.

In the fall of 1960, Caltech astronomers Thomas Matthews and Allan Sandage prepared to use the 200 inch telescope at Mount Palomar in California to make some observations of a radio source denoted 3C48 (the forty-eighth entry in the third "Cambridge catalogue" of radio sources). They were interested in what kind of visible light this source might be emitting, so on the night of 26 September 1960 they took a photographic plate of the area of sky around 3C48. Conventional wisdom at the time told them that they would find a cluster of galaxies at the location of the radio source, but this was nothing like what they saw. Instead, as far as anyone could tell by looking at the photographic plate, the object was a star. Yet it was like no other star seen up to then, for subsequent observations during October and November of that year and periodically throughout 1961 showed that its spectrum of colors was highly unusual, and that its brightness or luminosity varied widely and rapidly, sometimes over periods as brief as 15 minutes. This was a new addition to the astronomical family, and it needed a special name. It was a powerful radio source, yet it looked "stellar" or starlike (ordinary stars are not strong radio sources); on the other hand, because of its spectrum and variability it was not quite a star, it was only "quasi" stellar. Hence the name quasistellar radio source or "quasar" was soon applied to this object and to others like it.

The discovery of quasars brought general relativity to the attention of astronomers. The reason was an energy crisis of truly cosmic proportions. Within a few years after the discovery of 3C48, it was found that it and other quasars like it were among the most distant objects in the universe. What the astronomers thought were unusual spectra were actually rather ordinary spectra in which all the features were shifted uniformly to the red end of the frequency spectrum. This meant that the quasars must be moving away from us at high speeds, 30 percent of the speed of light in the case of 3C48. The shift in wavelength to the red is a consequence of the expansion of the universe. For 3C48, for instance, the recession velocity corresponded to a distance of about six billion light years. Because the quasars were so distant, one would have expected them to be faint, yet they were very bright sources, both in visible light and in radio waves. Therefore, their intrinsic brightness or luminosity must be enormous. For 3C48, the numbers translated into a hundred times the brightness of our own galaxy.

This was the energy crisis: What could possibly be the source of such power? On cosmic scales the strongest force known is gravity, so it was suggested that the energy of super-strong gravitational fields could provide the answer. Furthermore, the source of this power had to be very compact, for the simple reason that for the source to vary in brightness coherently over a period of, say, one hour, it couldn't be much larger than the distance light can travel in one hour, in order for one side of the source to know what the other side is doing and thus to behave in unison.

Thus, one solution to the quasar energy crisis involved strong gravitational fields, meaning perhaps a huge concentration of mass, maybe millions of times the mass of the Sun, confined to a region of space smaller than a light hour, or about the diameter of the orbit of Jupiter. This represented a new collapsed state of matter that could only be described by the general theory of relativity.

But relativists and astronomers knew almost nothing about each other, worked on entirely different problems, were housed in entirely different departments within universities, and spoke different scientific languages. To remedy this, in June 1963 a small group of relativity researchers based in Texas sent invitations to astronomers and general relativists around the world to attend a conference on a proposed new discipline, to be called relativistic astrophysics. The First Texas Symposium on Relativistic Astrophysics was held in Dallas on 16–18 December 1963 (page 5). The atmosphere was a mixture of excitement, because of the potential for solving an important problem by bringing these communities together, and grief, because of the assassination of President John Kennedy in that city just three and a half weeks earlier. Indeed, Texas Governor John Connelly, his arm still in a sling from having taken one of the assassin's bullets, opened the conference and welcomed the participants. There were 300 attendees, of whom roughly 240 were astronomers or astrophysicists, and 60 were relativity researchers. The latter number represented almost all of the world's general relativists at the time. The only ones missing were relativists from Eastern Europe and the Soviet Union, this being the middle of the Cold War.

The problem of quasars took center stage, and the leading models to resolve the energy crisis involved the collapse of great masses to the Schwarzschild "singularity." But what was collapsing? William Fowler of Caltech and Fred Hoyle of Cambridge University proposed the collapse of a supermassive star, perhaps millions of times the mass of the Sun. Cornell astronomer Thomas Gold suggested the collapse of an enormous and dense cluster of stars. John Wheeler and his post-docs and students presented papers on the collapse of compact objects such as neutron stars. It is interesting in retrospect to notice that all the models discussed were about the collapse process, while the final Schwarzschild singularity, as it was still called, played no essential role. The concept of the black hole as a standalone object was still poorly understood in 1963, and it would be several decades before supermassive black holes would be identified as the "central engine" for the power of quasars.

There were very few papers devoted purely to general relativity and its consequences. One, given by a young Ph.D. student of Wheeler named Kip Thorne, was on a toroidal, or donut-shaped, configuration of pure electromagnetic fields held together by gravity, a rather esoteric topic. The other was a mathematical paper by a physicist from New Zealand named Roy P. Kerr, who was working at the University of Texas at Austin. He had been using a variety of sophisticated mathematical techniques that exploited symmetry principles to look for new exact solutions of Einstein's equations. The solution he obtained was expressed in a fairly obscure system of mathematical variables, and so when he gave his talk he must have seemed like a visitor from another planet to the astronomers, who had not yet learned how to comprehend relativistic jargon. But during the question period after his talk, the Greek relativist Achilles Papapetrou admonished the audience to pay attention to this young man's solution, because he had a feeling it would one day prove to be important.

Indeed, Kerr's solution was soon identified as the exact solution for a *rotating* black hole and became the basis for all of modern black hole physics. Schwarzschild's solution was for the special case of a non-rotating black hole, but, since almost everything in the universe—planets, stars, galaxies—rotates, the Kerr solution would prove to be more physically relevant.

The fact that astronomers and general relativists were being brought together to work together on these kinds of questions was exciting, although at first it had its amusing side. Several participants at that first Texas symposium tell of a general relativity theorist interrupting a lecture by an astronomer to ask what he meant by the "magnitude" of a star (magnitude is the astronomer's measure of the brightness of a star, an elementary concept taught in every freshman astronomy class), or of the astronomer asking the general relativist what the "Riemann tensor" was (the Riemann tensor is a measure of the curvature of spacetime, to the relativist an equally elementary concept). There were skeptics of this attempt to get the two fields to play together nicely. The MIT astrophysicist Philip Morrison proclaimed himself "interested but unpersuaded" that new physics would come out of collapse to the Schwarzschild radius, while Peter Bergmann admitted that he was "not very optimistic" that the play date between the two communities would amount to much any time soon.

But Tommy Gold had the last laugh during his banquet speech, declaring:

It was . . . [Fred] Hoyle's genius which produced the extremely attractive idea that . . . the relativists, with their sophisticated work, were not only magnificent cultural ornaments, but might actually be useful to science! Everyone is pleased: the relativists, who feel they are being appreciated, who are suddenly experts in a field they hardly knew existed; the astrophysicists, for having enlarged their domain, their empire, by the annexation of another subject, general relativity. It is all very pleasing, so let us all hope that it is right. What a shame it would be if we had to go and dismiss all the relativists again!

Soon, however, the practitioners of this new interdisciplinary field learned how to communicate with each other, so that by later Texas Symposia (the twenty-ninth was held in Cape Town, South Africa in 2017), it was not uncommon to find relativistic astrophysicists who were as knowledgable about the intricacies of curved spacetime as they were about the structure and evolution of stars or about the capabilities and limitations of X-ray telescopes.

From 1963 to 1974, many of the key physical and mathematical properties of black holes were established during a period of intense research by a score of theorists. They learned that, to an observer outside the horizon, the only feature of the black hole itself that is detectable is its gravitational field. All information about what went across the horizon either during the formation of the black hole or during its later life is lost. Any matter or radiation that remains outside the horizon, of course, is detectable. Far away from the black hole, this gravitational field is indistinguishable from the gravitational field of any object of the same mass and angular momentum, such as a star. However, to an observer close to the horizon, things can be very unusual. The deflection of light can be so large that light can be deflected by large angles, not just a few arcseconds. A light ray can even move on a circular orbit just outside the horizon, at 1.5 times the Schwarzschild radius, for a non-rotating black hole. For the Kerr solution, the rotation of the black hole produces the same effects of the dragging of inertial frames as those induced by the rotating Earth. These are the effects confirmed by Gravity Probe-B and the LAGEOS measurements (Chapter 4). But if the observer goes close enough to the horizon, near the equator, the dragging of spacetime becomes so strong that the observer will be dragged around bodily with the rotation of the hole, no matter how hard he blasts his rockets to try to avoid whirling around the body.

But instead of dwelling on the many unusual and remarkable properties of black holes, we will turn to the observational search for black holes, looking particularly for examples where tests of general relativity might be feasible.

While the discovery of quasars spurred interest in the role of general relativity in astrophysics, it would be several decades before the central role of black holes in the quasar phenomenon would be appreciated. Instead, the first serious candidate for an actual black hole in nature came in 1971, from the new field of X-ray astronomy.

The first astronomical X-rays from sources other than the Sun were discovered beginning in 1962, including a source called Cygnus X-1, the name denoting the first X-ray source found in the constellation Cygnus. By 1967 about thirty such sources were known, all detected using instruments placed on sounding rockets or balloons launched far above the Earth's absorbing atmosphere. However, X-ray astronomy made a giant leap into the mainstream of astronomy with the launch of the Uhuru orbiting X-ray satellite in December 1970. The name Uhuru, meaning "freedom" in Swahili, was given to the satellite because it was launched from a facility in Kenya on that country's independence day (NASA's official name was the typically boring "X-ray Explorer Satellite SAS-A"). During its three-year lifetime, Uhuru charted more than three hundred X-ray sources. Later orbiting X-ray satellites found many more sources, including ordinary stars, white dwarfs, neutron stars, galaxies and quasars, as well as a diffuse background of X-rays, reaching us from all directions.

Uhuru's examination of the X-rays from Cygnus X-1 gave two crucial pieces of information that led to the conclusion that a black hole was present. The first was the observation that the X-rays were variable in time in an irregular fashion, but on timescales as short as a third of a second. This meant that the region from which the X-rays originated had to be of the order of a third of a light second, or around 100,000 kilometers in size. This, in turn, implied that the object at the center of the X-ray emitting region had to be a very compact object, such as a white dwarf, a neutron star or a black hole, because a normal star, like our Sun, would have a diameter ten times too large. The second piece of information provided by Uhuru was an accurate enough position for the source in the sky to make it possible to locate a star, known as HDE 226868, at the same location. Examination of the spectrum of light from this star showed that it was in orbit about a companion. This was determined by looking at the Doppler shifts in its spectral lines, just as the orbits of binary pulsars are determined by looking at the Doppler shifts of their pulse periods (see Chapter 5), or as exoplanets are found using Doppler shifts of the spectra of their parent stars. The companion had to be the X-ray source.

You may be wondering exactly how a black hole meets up with a star in order to perform this dance, since after all, space is a very large and empty place. The standard scenario begins before the black hole was a black hole, back when it was a star orbiting a companion star. Even though our Sun is the only star in the solar system, as many as half the stars in our galaxy are in binary systems, orbiting around each other just as Earth orbits the Sun. After a long enough period of time one of these stars will run out of fuel to burn via thermonuclear reactions, and if it is massive enough it will collapse with an accompanying supernova explosion of its outer layers. For a relatively low-mass initial star, the explosion and collapse frequently produces a neutron star. This is the pathway that can lead to pulsars in binary systems, as we discussed in Chapter 5 (page 80). But if the initial star has a higher mass, the implosion of the core does not halt at the neutron star stage but proceeds all the way to a black hole. What is left then is a black hole and a star in a binary system. Subsequently, if the stellar companion itself is massive enough, it may also eventually undergo a supernova explosion and a core implosion, leading to a black hole companion. Such binary black hole systems will be lead characters in our story of gravitational wave detection in the next three chapters.

But back during the black hole–star phase, if the black hole is close enough to its companion star, its strong tidal gravitational force can distort the star into a shape somewhat like a teardrop (Figure 6.2). At the tip of the teardrop the force of attraction toward the black hole is stronger than that toward the star, and so gas migrates from the star toward the black hole. But the gas does not head right into the black hole, because the hole's orbital motion has carried it sideways a bit. So just as two ice skaters passing by each other quickly lock arms and begin a rapid spin around each other, the streaming gas is grabbed by the black hole's gravity and swirls around it in a gaseous disk. Because a ring of gas at a given distance from the black hole moves a little faster than a ring just outside it and a little slower than a ring just inside it, there is friction between adjacent rings of gas. This friction has two important consequences. It heats the gas to such high temperatures that the gas emits light all the way into the X-ray band. The friction also slows down

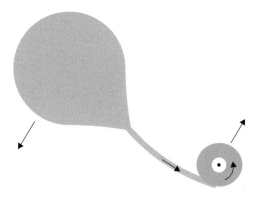

Fig 6.2 Accretion of gas from a companion star onto a black hole. The black hole is represented by the black dot, surrounded by an accretion disk of hot gas that can emit light in the X-ray band. Inside the inner edge of the disk (white region), gas can no longer be in a steady circular orbit but instead plunges directly into the black hole.

the rings of gas, causing them to spiral inward to the black hole. When the gas reaches a distance from the hole around three times the radius of the horizon, it can no longer maintain a steady circular orbit and it plunges toward the hole, crossing the event horizon and adding a bit to the mass of the hole. This inner edge of the disk is represented by the white region in Figure 6.2. A disk like this is called an "accretion disk" because the gas eventually is accreted by the black hole.

This model accounts for the main features of the X-ray source Cygnus X-1, and for many such X-ray sources discovered subsequently. In some cases, in addition to the gas torn from the stellar surface, the massive stellar companion may emit a strong stellar wind, much like the solar wind, but on a much more massive scale. Some of that gas can also find its way to the accretion disk around the black hole. This is thought to be true of HDE 226868, the companion star in Cygnus X-1.

But what makes us think that Cygnus X-1 involves a black hole? Could it not be a neutron star or a white dwarf? It is here that we combine general relativity with information on the orbital motion of the companion star to identify the compact object. From studying the spectrum of HDE 226868, astronomers concluded that it is of a type of star that typically has a mass between 20 and 40 solar masses. In order to induce the observed orbital motion of the star, the mass of the compact object must therefore be at least 10 solar masses. It cannot be a white dwarf, because, as we remarked in Chapter 5 (page 126), the maximum possible mass for a white dwarf is about 1.4 solar masses, the Chandrasekhar mass. This conclusion does not depend on general relativity, because white dwarfs are not very relativistic. What about a neutron star? General relativity plays an important role in the structure of neutron stars; nevertheless, relativists have determined a maximum possible mass for them as well, in this case about 3 solar masses, and by no means as large as 10 (the maximum *observed* mass of a neutron star is about 2.2 solar masses). Therefore, it is not a neutron star. The only object left that can be massive enough to induce the orbital wobble of the companion, yet is small enough in size to allow the short-term X-ray fluctuations, is a black hole. Even though this argument is somewhat indirect, it has stood up to further observations of the system, as well as to attempts to propose alternate models that do not invoke black holes. Still, to be safe, scientists customarily call these black hole *candidates*.

Many other black hole candidates have been discovered in X-ray binary systems. Interestingly, as we will learn in Chapters 7 and 8, none of them is as massive as the 20 to 50 solar mass black holes detected using gravitational waves by LIGO and Virgo, and this has challenged astrophysicists to come up with scenarios that could produce such heavyweights. There are also numerous X-ray binaries containing neutron stars. In these systems the mass of the compact object is always smaller than the 3 solar mass limit imposed by general relativity, and in many cases the X-rays are pulsed, indicating that the infalling gas is interacting with the strong, rotating magnetic field of the underlying neutron star.

In addition, there is an intriguing subset of X-ray binaries that shed additional light on the difference between neutron stars and black holes. These are systems in which the rate of accretion of gas from the companion is very low, so that the gas in the disk is thin and friction is much weaker. As a result, the X-ray emission is very faint, but still detectable. However, when one looks at those systems in which the mass of the compact object is less than 3 solar masses, there is an additional

X-ray flux superimposed on the disk flux, while for every system where the compact object's mass is greater than 3 solar masses, there is only the feeble disk flux. Ramesh Narayan and his colleagues at Harvard University have suggested that in the high-mass systems, the gas reaches the inner edge of the disk and plunges into the black hole, emitting no additional radiation. But in the low-mass systems, the gas crashes onto the surface of the neutron star, heats up and emits the additional flux of X-rays. They have suggested that this is the first concrete evidence of general relativity's prediction of the existence of an event horizon. If this result holds up, it will be a test of a central prediction of Einstein's theory.

Other researchers are asking whether one can test general relativity by examining the details of the emission from such accretion disks, particularly variations with time and unique features in the spectra. After all, near the inner edge of the disk, the gas is orbiting in an extremely warped region of spacetime compared to that in the vicinity of the Sun. For a 10 solar mass black hole, the orbital period just before plunge is about 5 milliseconds and the gas is moving at half the speed of light. The radiation that the gas emits experiences strong Doppler shifts, extreme gravitational redshifts and strong deflections, with some rays encircling the hole a few times before heading out toward the observer. If the black hole is rotating, the dragging of inertial frames will induce a number of observable effects. The hope is to test whether the spacetime geometry around the compact object really is that of either the Schwarzschild solution or the Kerr solution. Unfortunately this is a very complex problem. One must somehow cleanly separate those phenomena arising from spacetime warpage from those arising from the complicated physics associated with the gas and radiation, sometimes called "dirty gastrophysics." This is currently an extremely active area of research, and may soon provide some remarkable new tests of general relativity.

The next place you might think to test general relativity using black holes is in quasars. There is widespread agreement that the large redshifts in the spectra of quasars indicate that they are moving away from us at large velocities, and that, according to the picture of the expanding universe, they are therefore at very great distances. The powerhouse of the quasar is believed to be the active and violent central nucleus of a galaxy. The idea that this nucleus involves a relativistic collapsed object has changed little since the first Texas symposium, but now a rotating, supermassive black hole itself is the central engine. The black hole may weigh 100 million solar masses; as large as this is, it may still be only a tenth of a percent of the total mass of the galaxy. The black hole is gobbling up stars and gas at a ferocious rate, perhaps as much as one solar mass of material per year. As the material approaches the hole, friction from collisions with other material heats it up to temperatures high enough to make it radiate the enormous power we see on Earth. The narrow jets of matter that can be seen shooting out at nearly the speed of light on opposite sides of many quasars are believed to be the product of an interaction between magnetic fields embedded in the accreting matter and the strong dragging of inertial frames by the rotating black hole (page 58). There is evidence that quasars were much more prevalent in the early universe than they are at present; as we look farther out in distance, we are also looking farther back in time because the light from the quasar takes a finite time to reach us. It has been found that the number of quasars peaks at a time corresponding to an age of the universe about one-third of its present age. This may be the result of the finite time needed to grow such massive black holes (a problem that is still not fully solved) and the fact that once the black hole has swept up the stars and gas from the core of the galaxy, the quasar phenomenon shuts off.

Although around 200,000 quasars have been found, the current view is that they are a small and temporary subset of a larger population, derived from the observation that essentially *all* massive galaxies contain massive black holes, and the fact that there are hundreds of billions of galaxies in the observable universe, far more than the number of quasars. These black hole masses range from 100,000 solar masses to the current world (or should we say, universe) record of 20 billion solar masses in the galaxy NGC 4889. It is still unclear exactly how these supermassive black holes form. One hypothesis is that these monsters form when many smaller black holes merge over the aeons of time. If enough small black holes form early enough in the universe, then they will be gravitationally attracted to each other and will merge. This scenario is aided by astrophysical mechanisms that guarantee that heavy objects tend to sink toward the center of galaxies, thus increasing the chances of mergers, as well as by the fact that mergers of galaxies themselves were rather common in the early universe.

With all these massive black holes around, you might expect a plethora of tests of general relativity. But just as with black holes in binary systems, the complications of gastrophysics make the problem hard, although this is also an area of current research. It turns out, however, that there is one supermassive black hole that may be the perfect laboratory for testing Einstein's theory. All you have to do is to . . . look up in the (southern) sky! It's close! It's clean! It's Sagittarius A*!

We will devote the rest of this chapter to the story of this remarkable, massive black hole, sitting smack dab in the center of our own Milky Way. The story begins with Karl Jansky (1905–1950), the pioneer of radio astronomy, whom we met briefly in Chapter 3.

Jansky was born in the Territory of Oklahoma to parents of Czech and French–English descent. He finished his undergraduate physics degree at the University of Wisconsin in 1927 and then moved to New Jersey to work for Bell Telephone Laboratory. At the time, the company wanted to investigate the use of electromagnetic waves with a short wavelength (of about 10 meters) in trans-Atlantic telephone services. In 1931 Jansky was tasked with studying what else could produce such waves on Earth and interfere with communication signals. To do this, he built a radio antenna designed to detect waves with a wavelength of about 15 meters (with a corresponding frequency of 20 megahertz), and mounted it on a large turntable. Jansky's "carousel" was about 30 meters wide and stood about 6 meters tall, allowing him to rotate the antenna and pinpoint the direction of any signals he detected.

Over several months, he collected data. The main sources of radio static were thunderstorms, but in addition to weather effects, roughly once per day his antenna also detected a faint but steady radio "hiss" of unknown origin. Whenever a scientist detects a signal that repeats once per day the usual suspect is the Sun, and Jansky at first reasoned that he was recording radio waves from the Sun.

Upon further study, however, Jansky realized that the signal repeated once every 23 hours and 56 minutes and not every 24 hours. The latter is the time it takes the Earth to complete a full rotation so that the Sun appears in the same position in the sky. This period is called the solar day. But 23 hours and 56 minutes is the time it takes the Earth to complete a full rotation so that the *stars* appear in the same position in the night sky. This is called the sidereal day. The roughly 4 minute difference is due to Earth's motion around the Sun, which of course affects when the Sun rises but has no effect on when the stars appear in the night sky. If the signal Jansky had detected had something to do with the Sun, then it should have repeated with the solar day and not the sidereal day. The data indicated an origin far outside the solar system.

By carefully rotating his antenna and taking many more months of data, Jansky was able to show that the signal was strongest in the direction of the center of the Milky Way galaxy. This coincides with the direction of the Sagittarius constellation, near a feature denoted Sagittarius A by astronomers. He published a paper on his discovery in the *Proceedings of the Institute of Radio Engineers* in 1933.

A *New York Times* article entitled "New Radio Waves Traced to Centre of the Milky Way" catapulted Jansky to brief public stardom. But despite this, he could not convince other astronomers that there was important science in this "star noise" he had detected. It did not help either that the United States was going through the Great Depression in the early 1930s, followed by World War II. The field of astronomy would eventually recognize Jansky as the father of radio astronomy and name a unit of radio flux, the jansky, after him, but only after his death in 1950 from a heart condition.

For decades, Jansky's mysterious radio source in the Sagittarius constellation remained mostly unexplored, until in 1974 astronomers Bruce Balick and Robert Brown used radio interferometry to explore the region. Recall from Chapter 3 (page 39) that combining pairs or groups of radio telescopes can enable pinpointing the direction of a radio source with very high precision. In addition, this technique can resolve the size and shape of such sources with good resolution. Working at the National Radio Astronomy Observatory, Balick and Brown tried to resolve the patch of the sky from which Jansky had earlier detected radio waves. To their surprise, they found that most of the emission was confined to a very small area in the sky that was coincident with the Galactic Center. The size was about a tenth of an arcsecond as seen from Earth, or about 800 astronomical units at the source (later observations would narrow the size down to 50 microarcseconds, or about half an astronomical unit).

Balick invented the name Sagittarius A* for the radio source in 1982, arguing that in quantum mechanics the excited states of atoms are sometimes denoted with an asterisk, and this radio source was indeed very "exciting." Other names were later proposed for Sagittarius A*, but none of them stuck, and a standard abbreviation Sgr A* (pronounced "saj-ay-star") was soon adopted.

The fact that the radio emission was coming from such a compact region was a hint that a black hole might be there, but confirmation was difficult to come by. It was impossible to observe the region using optical telescopes, because of enormous bands of dust that lie between the solar system and the center of the galaxy, which absorb light in the visible band. However, in a band of wavelengths just beyond the red end of the visible spectrum, called the near infrared, light passes right through the dust, making the Galactic Center "visible" (albeit not by eye, but using special infrared sensors that had been developed to enable this branch of astronomy). Numerous astronomers trained their infrared telescopes on the Galactic Center to try to see what was going on there.

Two teams in particular took advantage of the latest advances in telescope technology. These included the ability to do interferometry at infrared wavelengths, extending a method that had been routine at radio wavelengths (see Chapter 3). Another advancement was a technique called "adaptive optics," whereby information about disturbances in the Earth's atmosphere is used to alter the shape of the mirrors of the telescopes in order to achieve the sharpest images. They also had the advantage of working at dry, high-altitude sites. Being high and dry is important because water vapor absorbs near-infrared light. One group, based at the Max Planck Institute for Extraterrestrial Physics in Garching, Germany, headed by Reinhard Genzel, used the Very Large Telescope Interferometer. This is an array of four instruments located on a mountaintop in Chile at about 8,600 feet above sea level, some 1,200 kilometers north of Santiago. The other group, led by Andrea Ghez of the University of California, Los Angeles, used the two telescopes of the 13,000 foot altitude Keck Observatory near the peak of the extinct Mauna Kea volcano in Hawaii.

But when they looked near the location of Sgr A* they saw something astounding. Stars! You might think that this should not be a big deal, since stars are what astronomers are usually paid to see. But these were neither foreground stars, between us and the Galactic Center, nor background stars, on the far side of the central region. Such stars are easy to identify and account for. These stars seemed to be in the close neighborhood of the Sgr A* object itself. The spectra of the light they emitted showed them to be of a class of massive, cool, young stars known as S-stars, around ten times more massive than our Sun. Accordingly, they gave the stars the highly original names S1, S2, S3, and so on (the UCLA group called them SO-1, SO-2, and so on; in 20 years the two groups have sadly been unable to agree on a common set of names). Ghez called this the "paradox of youth," because these stars could not possibly have formed there from the usual collapse of a large cloud of gas and dust, the way most stars such as our Sun formed. This is because the gravitational field of the central object would have disrupted the cloud before the star could form. So where did they come from and how did they get so close to the central object?

Even more interesting was that within a few years of observing the S-stars, the astronomers could see them move! The Munich group reported the first detections of motion in 1996 and the UCLA group followed two years later. There is a reason why ancient astronomers referred to the "fixed stars." They are so far away that it's nearly impossible to see them move. Even in the modern era, it takes careful long-term monitoring of stars with the utmost precision to detect their transverse motion, and even then it works only for stars in our immediate solar neighborhood (detecting their motion along the line of sight using the Doppler shift is much easier, of course). To see a star change its sky position at the Galactic Center after only a few years implied that the star was moving extraordinarily fast. Very soon it was realized that these motions were

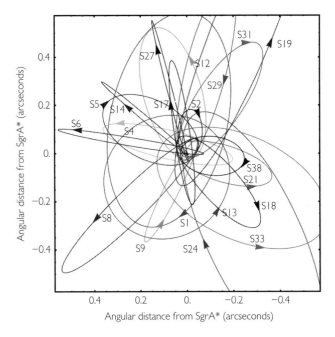

Fig 6.3 Orbits of selected S-stars around the Galactic Center black hole Sgr A*. The black hole is located at the exact center of the graph, with a diameter about 10,000 times smaller than the distance between tick marks at 0 and 0.2 on the graph.

not random, they were orbital. These stars were in orbit around the spot that had been established for Sgr A*.

The star S2 was particularly important, because it had a relatively short orbital period of around 16 years, and was in a highly elliptical orbit (see Figure 6.3). By 2002, S2 had reached its pericenter, or point of closest approach to Sgr A*, and there was data covering more than half of its orbit.[1] With this information, the teams were able to use

---

[1] Pericenter is the generic term denoting the closest approach in an eccentric orbit. For specific systems there are suitable variants, such as perihelion for orbits about the Sun, perigee for the Earth, perijove for Jupiter, periastron for binary stars and so on. The relativity community has not managed to come up with a good term for orbits around black holes. "Periholion" doesn't thrill the community. Our colleague Scott Hughes at MIT has proposed "peribothros," using the ancient Greek word βόθροσ for hole, but Greek colleagues have pointed out that in modern Greek this word has a different (curse-word) meaning. We invite readers to send suggestions.

Newton's theory of gravity to determine that the point about which S2 was orbiting, whatever it might be, had to be several million times more massive than the Sun. Einstein's theory is not needed here because the closest approach of S2 to Sgr A* is still distant enough that relativistic effects are unimportant for this calculation. A material object with such a mass, such as a hypothetical supermassive star or a dense cluster of stars, would have been visible in multiple bands. As we have seen, radio measurements had confined Sgr A* to a region several hundred astronomical units in radius. A tremendously massive object confined to such a very small region of space, that emits almost no light, has to be a black hole. Genzel's team announced this conclusion in *Nature* in October 2002, and Ghez's team made a similar announcement in early 2003. Improved measurements indicate that what is now called "the black hole Sgr A*" has the mass of 4.3 million suns.

Figure 6.3 shows the orbits of twenty S-stars that have been inferred from the data (over a hundred stars are currently being monitored). The location of Sgr A* is at the very center of the graph, where the lines corresponding to zero angle cross (you may be able to spot the dot just inside the very elongated orbit of S14). A black hole of its mass would be roughly 25 million kilometers in diameter, about 18 times the size of the Sun. The Galactic Center is approximately 26,000 light years away from us, which translates to about one and a half billion times the distance from the Earth to the Sun. At that distance, Sgr A* is about 20 microarcseconds, or 0.00002 arcseconds, in angular diameter. For comparison, this is the size of two American quarters on the Moon as seen from Earth. On Figure 6.3, Sgr A* would be a dot with a width about 10,000 times smaller than the distance between the tick marks at zero and 0.2 arcseconds at the bottom of the figure, much, much smaller than the dot shown.

Sgr A* is a hot playground for testing general relativity. Many of the tests we described earlier in this book can be repeated with careful observations of the stars in its vicinity. A prime example is the recent measurements of the gravitational redshift with S2, which we mentioned at the end of Chapter 2. Because it revolves around the black hole on a highly elliptical orbit, it reaches a mere 120 astronomical units at its pericenter, with a velocity of roughly 7,650 kilometers per second. Therefore, the light that S2 emits there is redshifted relative to the light it emits later in the orbit, because of both the special relativistic time dilation and the gravitational redshift effects. This redshift was measured in 2018 by Genzel's team and confirmed by Ghez's team in 2019, thus verifying general relativity's prediction one more time, but this time close to a black hole.

If this test of Einstein's theory sounds familiar, it is because we already encountered a similar test back on page 17, when we described the experiments conducted by Pound and Rebka at Harvard University in 1960. In those experiments, Pound and Rebka had control over the light source emitted (gamma rays with a narrow wavelength produced in the decay of an unstable isotope of iron), as well as over the height (74 feet for the Jefferson tower) between the emitter and receiver. What they did not have control over was gravity, which for Earth is so weak that the predicted shift in the frequency of light was a mere two parts in a thousand trillion. For S2, on the other hand, the predicted shift is about six parts in ten thousand, a much larger effect because of the much greater warpage of time near Sgr A*.

We began this book with a test of general relativity using Sgr A*, one of the events of "that very good summer" of 2017. Using orbital data from S2 and S38, a star with an orbital period of 19 years, Ghez's team searched for a specific deviation from the normal inverse square law that Newtonian gravity predicts for the gravitational force between bodies. General relativity also predicts the same law as a good approximation when the bodies are not too close to each other, which is the case here. But in some alternative theories to general relativity there could be an additional force, which could be attractive or repulsive, depending on the theory, and which falls off more rapidly with increasing distance than the inverse of the square of the distance. Since both S2 and S38 are on very elliptical orbits, they sample gravity over a wide range of distances, from the distance at the pericenter to almost ten times that distance. Thus their orbits were especially sensitive to any change in the force with distance.

No such anomaly was found. This was the first test of general relativity involving orbits around a black hole.

The elliptical nature of S2's orbit also allows for another classical test of general relativity with Sgr A*. As we saw in Chapter 5 when discussing the orbits of binary pulsars, curved spacetime induces a precession of elliptical orbits, leading to the famous perihelion advance of Mercury and the periastron advances of binary pulsars. The close proximity of S2 to Sgr A* at pericenter and the black hole's large mass lead to a precession of S2's orbit at a rate of 0.2 degrees per orbit, or about three-quarters of a minute of arc per year. On 19 May 2018, S2 passed through another pericenter, and many measurements of the orbit were made during that critical period when the effects of general relativity are the strongest. These observations should soon resolve this effect, and we expect a test of general relativity by roughly 2020.

But other tests of Einstein's theory with Sgr A* are possible if we find stars inside the orbit of S2. So far, no other stars have been detected closer to Sgr A*, both because of limitations of the telescopes and also because S2 is so bright and so close to the black hole that it makes it difficult to detect fainter companions between the two. If a companion were observed with the same orbital eccentricity as S2 but about twenty times closer to Sgr A*, it would be possible to measure the precession of the orbit due to the dragging of inertial frames. This frame dragging or Lense–Thirring precession is the same effect that we discussed in Chapter 4, when describing the measurements of the precessions of the orbits of the LAGEOS satellites. This would make it possible to measure the rate of rotation, or "spin," of the black hole.

Such a measurement is very important, for two reasons. The first is that the spin of a black hole of a given mass can range from zero for a non-rotating black hole, corresponding to the Schwarzschild solution, to a maximum value, corresponding to an extreme limit of the Kerr solution. The spin cannot exceed that maximum value, for if it did, the body would not be a black hole, but instead would be something called a "naked singularity," a bizarre object that physicists find so horrifying that they are sure that nature would never let them exist. The second reason is that a spin measurement would give hints as to how the black hole formed and grew to its large mass. If it was by the merger of two pre-existing smaller black holes, it would likely have a rather large spin, just as two ice dancers who pull together in a hug at the end of a dance are rotating quite fast. But if it was by the steady accretion of stars and gas falling across the event horizon from random directions, then its final spin might be rather low, since the matter absorbed would spin the hole up as many times as it would spin it down.

If stars even closer to Sgr A* were to be detected and tracked, then it might be possible to perform a test of the underlying assumptions of what is called the "no-hair" theorems of black holes. As theorists began to understand the full implications of the Kerr solution during the 1960s and 70s, they came to a startling realization. This solution was the *only* possible solution in Einstein's theory for a quiescent black hole sitting in otherwise empty space, with the Schwarzschild solution being the limiting case of no rotation. And all the details of the external gravitational field of a black hole depend on only two quantities: its mass and spin. If you have two black holes of the same mass and rotation rate, and one was formed from the collapse of gas while the other was formed from the collapse of a huge cloud of Toyota pickup trucks, the external gravity will be identical. This is very different from, say, the Earth, whose external gravity field depends on the rotation of its molten core, the rigidity of its crust, the peaks of mountains and the depths of valleys. Recall from Chapter 4 that the Earth's field has been measured in exquisite detail by orbiting satellites like GRACE.

Pondering this remarkable property of black holes, John Wheeler coined the phrase "black holes have no hair." He imagined that if you found yourself in a room full of completely bald men, it might be hard to tell one man from another, in contrast to being in a room of men with full heads of hair. Wheeler's aphorism has been encoded into mathematical statements about the precise nature of the field around any rotating black hole. Therefore we can contemplate using a number of stars orbiting close to the black hole (they need to be close so that the relativistic effects on their orbits are detectable) to map out the gravitational field of the hole, the way GRACE satellites map the field of the Earth. But if these maps don't agree with the prediction of general relativity, then either the theory fails in the strong gravity regime of black holes, or Sgr A* is a heretofore unknown object, nothing like a black hole. Are there stars orbiting close enough? We don't know, but the teams of astronomers peering at the Galactic Center are on the hunt. The next few years should be very exciting, as S2 moves toward apocenter, the farthest point in its orbit from Sgr A*, perhaps revealing the presence of stars even closer to the black hole.

Seeing stars move around Sgr A* is not the same thing as seeing the black hole directly. Needless to say, we cannot see any signal that originates inside the black hole. Fortunately, we now know that the black hole is surrounded by a disk of gas that is accreting into the hole, and is radiating light. Some of this radiation is in the radio band, producing the waves detected by Balick and Brown. Follow-up observations have confirmed the presence of an accretion disk. But unlike the strong X-ray emitting accretion disks associated with black holes like Cygnus X-1, and unlike the incredibly luminous disks associated with quasars, the accretion disk around Sgr A* is a total wimp. Apparently, there just isn't enough gas migrating into the Galactic Center, either from ambient gas or from disrupted or exploded stars, to feed a luminous disk. Is it possible that Sgr A* was once a bright quasar, now reduced to a faint ember from a lack of fuel? This is an open question at the moment. But as faint as this emission is, the Max Planck team reported in 2018 that they had detected variations in the emission from the accretion disk that are consistent with motions around the black hole of hot spots within the gas. These hot blobs of gas are moving so fast, about 40 percent of the speed of light, that they must be orbiting right at the innermost edge of the disk, close to plunging into the black hole (see Figure 6.2). These blobs are moving in an extremely warped region of spacetime!

But what if we could take an actual "picture" of Sgr A*? What would we expect to see? The answer is complicated because the black hole can bend light in dramatic ways. Taking a picture of somebody with your smartphone is straightforward because the light rays move on straight lines from the subject to the lens of the phone's camera. Photographing things near a black hole is more like snapping a photo of yourself in front of the warped mirrors that you find in carnivals or fun houses.

Depending on where you stand, you could have a fat head (or even two heads) and a slim waist, or you could have a pea-sized head and an enormously fat waist.

Recalling our discussion of gravitational lensing (Chapter 3), a black hole acts like an extremely strong lens, warping and distending what you see, in the same manner as do fun-house mirrors. So if a black hole suddenly appeared in the night sky you would see a number of strange things. First, stars would appear to be pushed away from the black hole, as we illustrated in Figure 3.2. Because of gravitational lensing you might see multiple images of the same constellations, for example *two* Orion's belts, or two Big Dippers, as illustrated in Figure 3.10. You might also see stars or constellations that are not normally seen in that part of the sky, but that are actually behind you. In this case, some light rays from the star can pass by you from behind, swing around the black hole and then enter your camera, producing an image of a star that is not actually there (Figure 6.4). There will be a circular black disk in the center, which we would observe to be about 2.6 times the diameter of the black hole itself. This is not the actual black hole, but is about as close to it as you can see.

You may recall that earlier in the chapter we mentioned that light can actually orbit the black hole in a circular path at 1.5 times the Schwarzschild radius. This is called the "light ring" (Figure 6.5). The

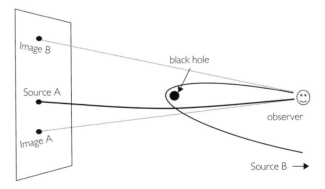

Fig 6.4 Strong lensing by a black hole. Not only can the black hole displace the image of a source (A), but it can also bend light from a source (B) *behind* the observer, producing an image (B) of a star that is normally not in that part of the sky.

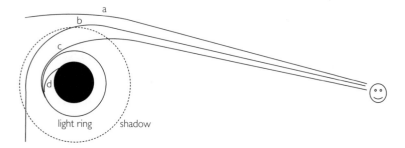

Fig 6.5 Black hole shadow. The black disk represents the black hole itself, and the solid circle is the "light ring," with a diameter about 1.5 times that of the black hole, where light rays can revolve around the hole on circular orbits. An observer detects a sequence of rays from the vicinity of a black hole, each ray passing closer than the previous. Ray "a" is deflected mildly; ray "b" passes closer and is deflected by around 90 degrees; ray "c" comes from just outside the light ring and is also strongly deflected before reaching the observer; ray "d" originates just inside the light ring, and has to cross the event horizon. Thus the observer cannot see any light rays from inside the dashed circle. This is the black hole's shadow, with an apparent diameter about 2.6 times that of the black hole's event horizon.

collection of all these rings is a sphere one and a half times the radius of the black hole horizon. Any light ray that enters this "light sphere" will spiral toward the event horizon and be lost (the light ring is not a horizon because the light ray could scatter off a passing atom and be deflected back out of the sphere). Any light ray that passes the black hole just outside the light sphere (path "c" in Figure 6.5) will reach your camera. But even after leaving the light sphere, the ray's path will still be bent quite a bit until it gets far from the hole. As a result, the black hole appears to a far-away observer as if it were casting a "shadow" about 2.6 times its diameter. In addition, you will see the black shadow outlined by a thin ring of faint light, coming from all the rays of light from all the stars and galaxies anywhere in the sky that just skim the light sphere, orbiting the black hole a number of times and then leaving the hole heading for your smartphone.

Similar distortions will occur if you photograph an accretion disk near a black hole. If you are looking at the disk from directly above (see the left panel of Figure 6.6), the image will look more or less like a disk, with

Fig 6.6 Schematic image of an accretion disk around a non-rotating black hole. Left: The disk seen face-on with the black hole shadow in the center. Middle: The disk is tilted relative to the line of sight. Some light rays from the top surface of the disk on the far side of the black hole are emitted vertically, then deflected by a large angle toward us in the warped spacetime of the black hole. The back side of the disk appears to be bent upward. Right: The disk is seen edge on. Some rays can be bent by very large angles by the black hole, so that we can simultaneously see both the top surface and the bottom surface of the disk *behind* the black hole.

an inner edge, where the gas plunges quickly toward the black hole with very little emission of light. The black center shown in the figure is the shadow of the black hole. If you are looking at the disk from an angle of, say, 45 degrees, then you will see the disk in front of the black hole's shadow, pretty much as expected, and part of the disk heading behind the shadow. But you will also see what looks like a bulge of the disk on the far side of the shadow (middle panel of Figure 6.6). This is not a physical bulge of the disk, but comes from light emitted upward from the top face of the disk that is bent by the black hole's gravity in a direction toward us. If you are looking at the disk almost perfectly edge on, you see a truly "fun-house" image (right panel of Figure 6.6). You will see the disk in front of the black hole shadow, but you will also see what looks like another disk circling the black hole. In the top part you are seeing the upper face of the disk behind the black hole, whose light starts out going upwards, but then is deflected by 90 degrees, heading toward you. In the bottom part you are seeing the underside of the disk behind the black hole. So in one picture you can simultaneously see both faces of the disk behind the black hole.

By discussing the non-rotating, or Schwarzschild, black hole, we have grossly oversimplified things to bring out the main points. In real life, we expect most black holes to be rotating, possibly at a substantial rate. If that is the case, then the dragging of inertial frames changes the shadow and distorts the images dramatically. The size of the light ring depends on the rotation rate of the black hole and on whether the light is going around the hole in the same direction as its rotation or in the opposite direction. And if the light is on an orbit that is tilted relative to the black hole's equator, then its motion can be very complicated indeed. For example, in the right panel of Figure 6.6, if the black hole is spinning with its rotation axis perpendicular to the accretion disk, then the bulges will be much more pronounced on the left side than on the right side of the hole. And the black hole's shadow will no longer be circular. Many of the calculations that lead to these pictures were done in the 1970s by James Bardeen, then at Yale University, and Jean-Pierre Luminet of the Observatory of Paris, during the period of intensive theoretical research into the properties of black holes.

In the movie *Interstellar* the image of Gargantua's accretion disk is a more detailed and correct representation of what we have shown very crudely in the right panel of Figure 6.6. The calculations done by Kip Thorne to generate that image took into account all of the gory details. However, some details had to be left on the cutting room floor. An example is various relativistic effects that make the light coming from the part of the disk that is approaching the observer much more intense than that coming from the receding part of the disk. Had these effects been included, the intense light would have completely saturated the image, leaving nothing but a bright spot. While astronomers and physicists love to observe and analyze such details, the film's director Christopher Nolan wanted a more pleasing image for his audience, so he asked Thorne to suppress those effects.

We are going to need a really good camera to take a picture of the disk around Sgr A*. After all, we already explained how Sgr A* subtends a tiny angle in the sky, a mere 20 microarcseconds, as viewed from Earth. Enter Very Long Baseline Interferometry (VLBI). In Chapter 3 we saw how connecting radio telescopes can lead to very precise measurements of directions of sources such as quasars (page 39). Combining the data from such telescopes in a specific way can also produce images of the source, given enough resolution.

The interferometer used by Balick and Brown only allowed them to limit the size of the region where Sgr A* is located, but not to resolve the source itself. To do better, you need longer baselines. In the late 1990s, astrophysicists Heino Falcke, Fulvio Melia and Eric Agol pointed out that if the baseline between telescopes is about the diameter of the Earth and the observing wavelength is about a millimeter, then the angular resolution at the Galactic Center would be around 20 microarcseconds, smaller than the diameter of the shadow of Sgr A* as seen from Earth. It was later realized that, with the same resolution, it would be possible to detect the shadow of the supermassive black hole in the center of the galaxy, Messier 87, generally called M87. Although, at a distance of 53 million light years, the galaxy is two thousand times farther away than Sgr A*, the black hole is 1,500 times more massive than Sgr A*, and therefore, as seen from Earth, its shadow is about the same size. Sgr A* and M87 seem to be the only two massive black holes so far that have just the right "Cinderella" combination of mass and distance to make this possible.

This inspired Sheperd Doeleman of MIT's Haystack Radio Observatory and Harvard's Center for Astrophysics to try to forge a collaboration of radio astronomers working at telescopes around the world to create an array with Earth-sized baselines. This would not be a simple task. The telescopes were operated by different agencies in different countries and had competing scientific priorities. Since each telescope would independently but simultaneously observe the Galactic Center, it was important that the instrumentation at each telescope be either identical or sufficiently similar that the data quality was the same. Each location had to have excellent atomic clocks so that the data could be time stamped accurately enough to permit merging the various data sets properly. The data from each observing session (thousands of terabytes worth of data) would be shipped to a central location for processing into images. It was also essential that the weather be excellent simultaneously at many different locations around the globe. Good luck with that!

As a proof of principle, Doeleman and colleagues managed in 2007 to use a triangle of short-wavelength radio telescopes in Hawaii, California and Arizona to detect *something* at the Galactic Center at a scale of the order of Sgr A*'s event horizon. This was the breakthrough needed to push ahead.

In 2012, Doeleman and colleagues formally kicked off the project, now called the Event Horizon Telescope. Currently, the array of telescopes in the collaboration numbers ten: two in the mainland USA, three in Chile, two in Hawaii, one in Spain, one in Mexico and one at the South Pole (see Figure 6.7). There are plans to add telescopes in Greenland, France and the USA. The initial full observation run occurred over ten days in April 2017, with eight observatories taking part. One obstacle to analyzing all the data was logistical: by the end of the run in April, winter had set in at the South Pole, and so the data files had to be placed in (literal) cold storage until December 2017, when they could be flown out. The data were copied and sent to four teams who analyzed them independently of each other, and with strict secrecy to guard against mistakes and to enable later checks and counter checks. Finally, on 10 April 2019, they announced that they had obtained an image of the shadow of M87, and that its size was consistent with the prediction of Einstein's theory. The data on Sgr A* is still being analysed; obtaining an image is more difficult, in part because the accretion rate of gas is on the low side, so it is not very bright, but also because the accretion is highly variable. More observations may be needed before they will be able to see Sgr A*'s shadow.

We all delighted in the breathtaking beauty of the pictures they produced; you would be hard pressed to find a newspaper anywhere in the world that did *not* have the image on its front page. But this is not the only (or even the main) reason for this major undertaking. The accretion disk around both black holes is not static, but changes with time as hot blobs of matter within the disk circle the black hole, thus changing the light emitted. As we already mentioned, such hot spots have already been detected around Sgr A* by the Galactic Center group at the Max Planck Institute. By combining a sequence of joint pictures, EHT hopes to produce a movie of how the accretion disk behaves as its inner edge

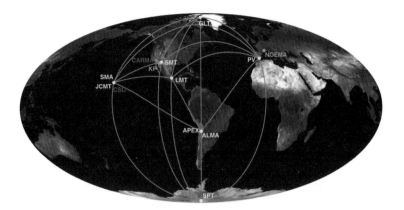

Fig 6.7 The Event Horizon Telescope. The radio telescopes of the project and the baselines joining them. Credit: Event Horizon Telescope Collaboration.

is swallowed by the black hole. This, in turn, will provide a detailed window into the physics and the dynamics of accretion disks, allowing astrophysicists to compare their models and predictions to actual data.

We may also be able to test general relativity with EHT. The idea is to test the black hole no-hair theorems much closer to the black hole than we could do using stars. In general relativity, the shadow cast by the black hole is a circle for the non-rotating hole, but is off center and somewhat flattened on one side if the black hole is rotating. Once you know the mass and angular momentum of the black hole, this shape is predicted precisely by general relativity, given a particular gastrophysical accretion disk model. But if rotating black holes are not described by the Kerr solution of general relativity, then this shadow will be different, perhaps more flattened or less flattened, perhaps shifted in a different way relative to the Kerr expectation. A precise enough observation of the shadow could therefore test general relativity.

Similarly, because the external gravitational field of the black hole is completely fixed by its mass and angular momentum, the response of an accretion disk and the behavior of the light it emits are predictable and fixed. The black hole no-hair theorems of general relativity leave no wiggle room. Of course, the light that EHT will observe will depend on the nitty-gritty details of the disk itself, such as its density, temperature and composition. But because, unlike its cousins in X-ray binaries and quasars, this disk around Sgr A* appears to be a relative weakling, there is reason to hope that there won't be too much dirty gastrophysics to complicate the interpretation of the observations. The same tests can be done with follow-up observations of M87.

We have come a long way from a time when Einstein and his contemporaries were sure that Schwarzschild's Massenpunkt solution would never happen in nature because of its strange singularity. Today we know that black holes exist, and they may soon provide remarkable new tests of Einstein's theory. But the black holes we have described in this chapter pretty much sit there being, well, black holes. What if we could detect two black holes colliding with each other?

CHAPTER 7

Gravitational Waves Detected At Last!

The room is small and windowless. At the front of the room stand five chairs, two large video panels and a podium displaying the logo of the NSF, the US National Science Foundation. The audience comes from around the world and includes scientists, government officials and reporters. They whisper in anticipation of a major announcement that the scientific rumor mill has been mongering for a couple of months. At 10:30 a.m. on Thursday 11 February 2016, NSF Director France Cordova welcomes the audience to the National Press Club in Washington DC.

Two thousand miles away, in the small town of Bozeman, Montana, Nico and twenty people sit at a table in a small room of the eXtreme Gravity Institute at Montana State University. This room has windows with a beautiful view of the mountains, but nobody is paying attention to the scenery. All eyes are focused on the television in the front of the room. The screen has a live internet stream from the National Press Club. Nobody pays attention to the celebratory cake waiting on the center of the table.

Five hundred miles to the south, in Aspen, Colorado, Cliff and eighty physicists and astronomers watch the same feed in an auditorium of the Aspen Center for Physics. They are participating in a workshop on stars and gas at the centers of galaxies, but the day's schedule has been pushed back by two hours so that everybody can watch this event.

After briefly extolling the NSF's commitment to funding cutting-edge research in fundamental science, Cordova sits down and the man sitting next to her approaches the podium. He is tall, middle aged, with graying hair, wearing a blue suit, with a blue shirt and a paisley tie. His tired eyes reveal that the past few months have seen very, very long hours. He places some notes on the podium.

"Ladies and gentlemen," he says. "We have detected gravitational waves. We did it!" he exclaims, and the audience bursts into applause. The twenty people in the room in Bozeman applaud; the audience in Aspen applauds. At institutes and universities around the world, scientists of all stripes break out in applause. David Reitze, the Director of the Laser Interferometer Gravitational-Wave Observatory, or LIGO for short, has just announced the most important scientific discovery of the twenty-first century (at least so far).

Around 1.2 billion years ago, in a very distant galaxy, two black holes crashed against each other. Each black hole was roughly thirty times more massive than our Sun, but in actual size was only about as big as Albania or Haiti. They were circling around each other at roughly half the speed of light, locked by gravity in a fatal dance, when they merged to form a single black hole. The event created ripples in the fabric of space and time that traveled outward in all directions at the speed of light. On 14 September 2015, those same spacetime waves finally arrived at the Earth, passed through the LIGO instruments and produced an unmistakable gravitational wave reading. This was the event that David Reitze had just announced to the world.

Within hours, congratulations poured in, from Stephen Hawking, from President Barack Obama, from leaders of the CERN accelerator center in Geneva. Twenty months later, a remarkably short time for the normally glacial Swedish Academies, the 2017 Nobel Prize in Physics was awarded to three of the founders of LIGO: Rainer Weiss, Kip Thorne and Barry Barish. "Gravitational wave astronomy" became an "official" field, hailed even by some of the astronomers who once lobbied against LIGO.

Gravitational waves were not always so in vogue.

At one point, Einstein himself thought he had proven that gravitational waves were not real! The definitive proof by theorists that they are real would not be achieved until the late 1950s, and the first experimental evidence that they exist would come in 1979, as we saw in Chapter 5. A 1969 claim by physicist Joseph Weber to have actually detected the elusive waves would soon be undone by the failure of other scientists to replicate his results. The story of gravitational waves is rich in science, of course, but is also a story of human personalities and foibles, of debates and controversies, and of big science, politics and money. It is a hundred-year-long saga that starts with a botched paper by Einstein himself.

In May 1916, Einstein published a major review article on his general theory of relativity that pulled together all the bits and pieces of the short papers that he had presented at the Prussian Academy of Sciences the previous November into a coherent exposition. He then immediately began to work on gravitational waves.

Einstein was a devotee of James Clerk Maxwell (1831–1879), the Scottish physicist who in 1867 united the seemingly disparate phenomena of electricity and magnetism into a single framework, known as electromagnetism. Maxwell's equations are still a central ingredient of modern physical science, from electrical engineering to high-energy particle physics. A deep understanding of Maxwell's theory underlies the technology in our most beloved devices, such as televisions, cellphones and laptops. Maxwell's equations are at the core of physics and engineering education today, and that was also the case in the late nineteenth century when Einstein was a student.

Maxwell's key insight was that electricity and magnetism could be understood through the idea of an electromagnetic *field*, a physical quantity that encodes information about the force exerted on a charged object anywhere in space. Even if you don't realize it, you have probably been exposed to the concept of a field before. You may have observed the way iron filings on a sheet of paper array themselves to display the field of the magnet under the paper. You know that Earth's magnetic field helps to protect us from the harmful energetic particles streaming from the Sun and creates the aurora borealis and the aurora australis. And you have heard of the gravitational field of the Earth, which is responsible for the force that allegedly caused the famous apple to fall on Newton's head, and that also holds the Moon in its orbit.

In addition, Maxwell showed that his equations had solutions in which the electric and magnetic fields oscillate, feeding off each other to produce a wave that travels at the same speed as light. He suggested that these waves *were* light, an idea that was confirmed experimentally by Heinrich Hertz in Germany in 1887.

Several scientists, including Hendrik Lorentz in the Netherlands and Henri Poincaré in France, began to wonder well before Einstein whether there could also be waves of gravity itself, simply by analogy with Maxwell's electromagnetic waves. Furthermore, because Einstein's special theory of relativity said that nothing could travel faster than light, it seemed logical that gravity should not be instantaneous. The effects of gravity should travel with a finite speed, and this speed should not be greater than that of light. But speculate was the best they could do, because they didn't have an actual theory of gravity to work with.

But in 1916 Einstein had an actual theory, and he set out, in the spirit of his hero Maxwell, to see if his equations had solutions that would resemble waves. He completed the calculations and published the result in June 1916. Unfortunately, the paper was full of what could charitably be called "bone-headed" mathematical and conceptual errors. Einstein's colleague, Norwegian physicist Gunnar Nordström, helped Einstein find and correct the mistakes, and Einstein published a second gravitational wave paper in 1918.

Einstein showed that a varying system, such as a dumbbell spinning about an axis perpendicular to its handle (Figure 7.1), will emit gravitational waves that travel at the speed of light. He also found that the waves carry energy away from the rotating dumbbell, just as light waves carry energy away from a light source. As we saw in Chapter 5, this loss of energy in a binary pulsar system is what Hulse and Taylor measured, thereby verifying, albeit indirectly, that gravitational waves exist. It's the same loss of energy that brought the two black holes to their final embrace, emitting the burst of gravitational waves that LIGO detected.

But there were aspects of his gravitational wave solution that Einstein didn't fully understand. In Maxwell's theory there are two solutions for electromagnetic waves. For example, if you have a wave propagating horizontally in the laboratory, one solution could have the oscillating electric

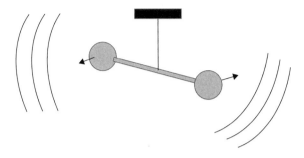

Fig 7.1 Einstein's spinning dumbbell generating gravitational waves.

field pointing vertically (it is always perpendicular to the direction of propagation); the other solution would then have it pointing horizontally but still perpendicular to the direction of propagation. These two cases are called the "modes of polarization" of the electromagnetic wave, and a general light wave consists of a combination of the two modes. A charged particle encountering the electric field of the first mode would move up and down, just as a ball moves up and down on a water wave at the beach. This is in contrast to sound waves, where the motion of the molecules of the medium which carries the sound is always *along* the direction of propagation of the waves. The concept of polarization is exploited for example in polarized sunglasses, which are designed to block one mode of polarization preferentially. They are most useful at protecting your eyes when the mode that is blocked is the one that is dominant when light scatters off the pavement in front of your car or off the water at the beach.

In the case of gravity, Einstein also found two modes of polarization of the gravitational waves (we'll get to what those modes look like shortly), and those modes traveled with the same speed as light. But there were additional solutions to the equations whose meaning was not so clear, and to make matters worse, the speed of these modes was not fixed by the equations. In a 1922 paper, Eddington analyzed Einstein's gravitational waves carefully. He pointed out that Einstein had made a small calculational error in his 1918 paper, making his formula for the energy lost off by a factor of two. He also observed that the additional modes of Einstein were probably not physically real, but instead might be waves in the coordinates used to describe the problem. He made the dismissive remark that "the only speed of propagation relevant to them is the speed of thought."

What does the statement "waves of the coordinate system" mean? It is a tenet of general relativity that you can label points in space and time in essentially any way you like and measurements made of the physical phenomena using devices such as clocks or telescopes or laser beams will be the same. Let us describe a simple example, not in spacetime but on a two-dimensional surface in a suitable city like Manhattan in New York City, where a very nice Starbucks store is located. There are a number of ways to tell your friend where the Starbucks is. The city itself has a grid of numbered avenues, say First, Second, Third Avenues pointing in one direction, and of numbered streets, say 65th, 66th, 67th Streets, pointing in the perpendicular direction. Many US cities are laid out in this regular way. This grid provides a "coordinate system" for the city, and let's say the nice Starbucks is located at the corner of Third Avenue and East 66th Street. However, you could use a very different coordinate system to direct your friend. You could give her the GPS coordinates of the store, and she could get there using her smartphone. The GPS coordinates are based on a grid of lines of longitude and latitude, which are not in general aligned with the avenue–street grid. Starbucks is there with your latte waiting no matter what coordinate system you use.

Now consider a third coordinate system, defined by a grid of parallel ropes held taut by giants standing at the edge of the neighborhood where Starbucks is located. There are two sets of parallel ropes, one set perpendicular to the other, with the ropes in each direction labeled E1, E2, E3, etc., and N15, N16, N17, etc. In this system, Starbucks might be located at E3–N17, i.e. at the intersection of rope #E3 in one direction and rope #N17 in the other direction. Your friend could just as easily get to Starbucks using this system. Now suppose that the giants start jiggling the ropes so that they all oscillate in some complicated manner

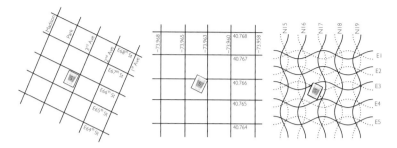

Fig 7.2 Coordinate systems for finding Starbucks in upper east side Manhattan. Left: System based on the grid of streets. Middle: System based on GPS coordinates. Right: System based on grid of numbered ropes being jiggled crazily by giants.

in the horizontal plane. In this coordinate system, the Starbucks store seems to be going crazy, now at E3–N17, now at E2.9–N16.8, then back to E3–N17, then E3.1–N17.2, and so on. Using this coordinate system to get your friend to Starbucks would be complicated (in fact the values of the coordinates would depend on her time of arrival), but it could be done. Yet the customers at Starbucks feel nothing, since after all, the store is not actually moving. They have no idea that the location of the store is sloshing back and forth when described using the giants' wavy coordinates.

This is the bottom line: coordinates are merely convenient labels of points and have no physical significance. In general relativity there are four coordinates, three for space and one for time, but the basic idea is the same. We can label points in spacetime using any convenient coordinates, but they will have no consequences for the physics that goes on, including the experiments we do using physical apparatus to measure the physical phenomena. Today, students in our general relativity courses learn this up front and eventually become comfortable with it.

But Einstein was not totally comfortable with it, even though he used the concept as the central guidepost for developing his theory. After all, he called it the "general" theory (as opposed to his 1905 "special" theory of relativity) because the theory was valid in "general" coordinate systems. Even Eddington, who understood general relativity at least as well as Einstein did, and probably better than any one else at that time, seemed slightly unsure what to make of these wavy coordinates. Today we think of their ruminations on the meaning of these modes as rather quaint, but we have the benefit of hindsight based on a century of research and teaching in the subject.

Still, Eddington's comment about some gravitational waves traveling at the "speed of thought" served to make the whole topic seem rather dubious, and it would be almost thirty-five years before the issue was fully settled.

The case for gravitational waves was not helped when, in 1936, Einstein tried to claim that they do not exist.

In 1933 Einstein moved to the United States to escape Nazi Germany, taking a job at the Institute for Advanced Studies, near Princeton University in New Jersey. This institute had been founded three years earlier to serve as a center for knowledge and discovery, and also became a refuge for intellectuals escaping Nazi oppression. In 1934 Einstein hired Nathan Rosen as his assistant, a physicist from Brooklyn, New York who had studied physics at MIT. An "assistant" in those days played the role of what today we would call a "postdoc," someone typically with a physics Ph.D. who acts as a research assistant to a senior scientist in a hands-on apprenticeship for a few years.

Einstein and Rosen set out to revisit Einstein's 1918 calculation to determine whether gravitational waves were truly real. Einstein's 1918 paper had used an *approximate* version of his theory to predict the existence of such waves, but they now wanted to determine whether the *exact* theory led to the same prediction. To their surprise, the exact theory seemed to predict the opposite! The solution they found was singular, or infinite, in certain parts of spacetime, just as Schwarzschild's "Massenpunkt" solution was singular at the Schwarzschild radius. Reasoning that singularities are unphysical, they concluded the entire solution also had to be unphysical. In 1936, Einstein and Rosen submitted their disproof of gravitational waves for publication in the *Physical Review* scientific journal.

Following a policy that had only recently been instituted, the editor of *Physical Review*, John Tate, sent the Einstein–Rosen paper to an external scientist to be reviewed. Given that one of the authors was Einstein,

Tate had some misgivings about this, but saw no reason to make an exception, especially considering the surprising claim made in the paper. The anonymous referee recommended that the paper not be accepted for publication without major corrections.

Today, peer review is the *norm* for publication in serious scientific journals, and it is an essential tool to ensure the validity of new results. Typically, a journal's editor will send a new submission to one or more experts for comments and critiques, and it is only once these experts reach a consensus for publication that the submission is accepted. But in 1936 this practice was not common at all and was almost unheard of in Europe, where Einstein was used to publishing papers. So when Einstein received Tate's reply with the anonymous referee's criticism, he was offended and withdrew his submission. He wrote to Tate that he had "sent the manuscript for publication," not to be disclosed to an anonymous expert. Einstein never submitted another paper to *Physical Review*.

If Einstein had actually read the report of the referee, our story would be quite different, since the referee had spotted a serious mistake in their paper. Instead, Einstein submitted the original paper, without any modifications, to the *Journal of the Franklin Institute*, a small journal published in Philadelphia, which accepted it without refereeing in 1937. By this point, Rosen had moved to the University of Kiev in present-day Ukraine, so Einstein had hired a new assistant, Leopold Infeld, a Polish physicist. Infeld arrived at the Institute for Advanced Studies around the time that the Einstein–Rosen paper was being accepted by the Franklin Institute journal, and Einstein was excited to talk with him about his new paper and the discovery that gravitational waves did not exist after all.

Infeld was initially skeptical. It was hard to believe that Einstein's theory, which resembled Maxwell's theory of electricity and magnetism in so many respects, did not have gravitational waves similar to Maxwell's waves. But Einstein was an eminence in physics, and he soon convinced Infeld that his argument was correct. Around this time, however, Howard Percy Robertson, a Princeton professor who had done work that would lay the foundations of general relativistic cosmology, was returning from a sabbatical at Caltech. When Infeld met Robertson, he told him about the Einstein–Rosen result, but Robertson dismissed it and, a few days later, showed Infeld exactly what the problem was. Once again it was the coordinates!

Robertson explained to Infeld that if one were to transform the solution that Einstein and Rosen had found to coordinates adapted to a cylinder, then the infinities that so troubled them would all be pushed to the axis of the cylinder, where the source of the gravitational waves should reside, and where their solution would not be applicable. Infeld was impressed that Robertson appeared to have solved the problem just based on their brief discussion.

The singularities found by Einstein and Rosen are called "coordinate singularities" and are today understood to be artifacts of the choice of coordinates and therefore have no effect on the physics. Even as mundane a place as the South Pole on the surface of the Earth is the location of a coordinate singularity. There, the latitude is −90 degrees, but there are an infinite number of possibilities for the longitude, since all the lines of longitude converge there. But standing at the scientific research station at the South Pole you would never know this, and in fact you could lay down a nice grid of streets on the ice and find Starbucks (if there were one there) just as easily as you could in Manhattan. As we discussed before, physics cannot depend on the coordinate choices one makes. Infeld rushed to tell Einstein about this, only to be told by Einstein that he had just reached the same conclusion on his own.

With input and advice from Infeld and Robertson, Einstein heavily revised the paper, already in galley proofs from the journal. He changed the title, added a new section on cylindrical gravitational waves, and altered the main conclusion: his exact theory of general relativity *did* predict the existence of (at least cylindrical) gravitational waves. In 2005, through some inventive detective work, our colleague Daniel Kennefick was given permission by the current editors of *Physical Review* to examine Tate's log book of submitted papers and confirmed that the anonymous referee of the Einstein–Rosen paper had been none other than Robertson himself!

Once again, sorting out what was real about gravitational waves proved to be a problem, and would not really be resolved for another twenty years. Recall that this period was the "low water mark" for the field of general relativity, when few scientists were interested in it or worked on it. But in the mid 1950s the beginnings of a rebirth in the subject began to take hold, leading to the great renaissance for general relativity during the 1960s described in Chapter 1. Science historians who study our field point to two conferences on general relativity as being particularly influential. The first was in Bern, Switzerland in 1955. The meeting was convened to celebrate the fiftieth anniversary of Einstein's "miracle year" when, while working as a patent clerk in that city, he developed special relativity and made groundbreaking discoveries in quantum mechanics and atomic physics. It also turned out to memorialize Einstein, who had died three months before the meeting. The second conference was in Chapel Hill, North Carolina in 1957.

In fact, these two meetings became so legendary that when the International Committee on General Relativity and Gravitation was formed a few years later and decided to organize a big "GR" conference every three years, they retroactively renamed these two meetings "GR0" and "GR1." In 2019, GR22 was held in Valencia, Spain.

Although gravitational radiation was not a major topic in either meeting, it was discussed. In Bern, Nathan Rosen reviewed cylindrical gravitational waves. In Chapel Hill, John Wheeler and Joseph Weber argued that, despite the rather unrealistic physical setup, cylindrical waves were physically measurable. And a 27-year-old British graduate student named Felix Pirani discussed his recent paper showing precisely how the waves would affect material particles and how you would measure the effects.

Imagine a ring of eight disks (hockey pucks, for example) arrayed in a circle on a table, free to slide across the table without any kind of friction (see the snapshot on the far left of the top panel of Figure 7.3). A gravitational wave passes vertically through the table (into the page of the figure). The strength or amplitude of the wave is a sine wave, starting at zero, growing to a maximum, returning to zero, then to a minimum and then returning to zero. According to Einstein's theory, the disks on opposite sides of the center of the circle are pushed apart in one direction while being pushed together in the perpendicular directions,

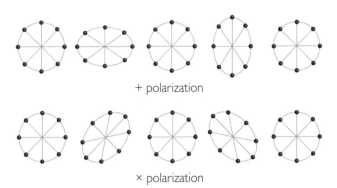

+ polarization

× polarization

Fig 7.3 Two modes of polarization of a gravitational wave propagating perpendicularly into the page. Snapshots show a set of eight disks on a frictionless surface. From left to right, the snapshots show the disks every quarter of a cycle. The top row displays the plus polarization; the bottom row shows the cross polarization.

as in the second snapshot in the top panel of Figure 7.3. As the wave amplitude passes through zero, the disks then return to the circular shape (third snapshot). As the amplitude goes negative, the disks are pushed together and apart in the opposite sense (fourth snapshot). Finally, after one complete cycle of the wave, the disks return to the circular shape (fifth snapshot). As we have already discussed, a general gravitational wave can contain two modes of polarization. The action of the second mode is shown in the bottom panel of Figure 7.3. It is the same as the first mode, except that the pushing together and apart is along the diagonals, at 45 degrees compared to the first mode. The two modes are conventionally called the "plus (+)" and "cross (×)" modes, because if you overlay the five snapshots in each case, the pattern reminds you of a plus sign or of a multiplication or cross-out sign.

Now, we have mentioned many times already that gravity is really due to the warping or distortion of spacetime, and you may have read many newspaper articles or internet stories describing gravitational waves as stretching and squeezing spacetime in order to produce the patterns shown in Figure 7.3. So it is fair to ask, doesn't the wave also stretch and squeeze the table so that the disks don't actually slide? The answer is that the response of an object to warping spacetime depends on what other forces are acting on it. Our disks have no forces acting on them in the horizontal direction (assuming zero friction) and so they respond "fully" to the spacetime distortions induced by the wave. On the other hand, the atoms that make up the table are being acted upon by the interatomic electric forces produced by all the surrounding atoms, and these forces are enormous compared to the force of gravity. So the distortion of the table is much, much, much smaller than the displacement of the disks. Thus, the disks will actually slide.

In fact, this was the basis of an ingenious argument made at the Chapel Hill conference by a "Mr. Smith" to underscore the reality of gravitational waves. Mr. Smith was actually the famous American physicist Richard Feynman, who had registered for the meeting under the pseudonym in part to indicate his low opinion about the status of general relativity research at the time. Drawing on Pirani's description of how objects would move, he pointed out that if you introduced a bit of friction onto the surface of the table, then as the disks slide back and forth, the table would heat up a bit. Therefore some energy would have been transferred from the gravitational waves to the tabletop, a clear and unambiguous sign of a physical effect. He actually used a pair of beads sliding on a rod in his discussion, and so his argument has gone down in physics history as Feynman's "sticky beads" proof of the reality of gravitational waves.

Pirani received his Ph.D. in 1958 under Hermann Bondi at King's College, London, and over the next few years he, Bondi and others would show in clear and unambiguous terms that gravitational waves are real, can be measured and carry energy away from their sources. The long period of uncertainty inspired by Eddington's jibe was over.

So we have answered one question: gravitational waves are physically real and measurable.

But an equally important question is: are there any waves to measure? Any decent physicist would say, let's generate the waves and then detect them. In 1887, Hertz had created electromagnetic waves using an electrical discharge, and then detected the effects of the waves at the other end of his lab. Could one take one of the dumbbells that Einstein imagined in his 1916 calculations and generate gravitational waves to be detected somewhere else? Unfortunately, it was not too hard to show that the waves produced by such a setup would be hopelessly weak, utterly undetectable by even the most far-fetched scheme. However, gravity depends on how much mass you have, and so perhaps one needs astronomical bodies to generate gravitational waves. But in the late 1950s, this did not seem very promising either. According to the astronomers, the universe was a very quiet place, where almost nothing happened. Planets revolved sedately around the Sun, stars hardly ever changed, and galaxies just sat there, moving slowly away from each other as the universe gradually expanded. To be fair, there were supernovae, stellar explosions seen, for example, in 1054, 1572 and 1604, but these were rare events, and it was not known how much if any gravitational radiation they emitted.

A third question is: if there were gravitational waves passing through the Earth, could one build a practical detector sensitive enough to sense them? The person who took on this challenge was Joseph P. Weber. To his many detractors Weber was a tragic figure, whose work was flawed and whose claims were roundly refuted. To others, he was the father of gravitational wave detection whose insights established many of the principles that enabled the successful detections using laser interferometers. His story illustrates the sometimes fitful and complex ways in which science advances, and is also a case study in how science works.

Weber was the son of Lithuanian Jews who settled in New Jersey and New York in the early 1900s. Born in 1919, he graduated from the US Naval Academy in 1940, served in World War II, and after the war led the Navy's electronic countermeasures section, retiring from the Navy as lieutenant commander.

In 1948, the University of Maryland appointed him as an engineering professor under the stipulation that he quickly earn a Ph.D. degree. Weber asked George Gamow, a physics professor at George Washington University in Washington DC, famous for his explanation of radioactivity through quantum physics, if he would be his Ph.D. advisor. Ironically, this was the same year that Gamow and his student Ralph Alpher had theoretically predicted the existence of the first light after the big bang, today called the cosmic microwave background (CMB) radiation. But instead of suggesting to Weber that he should work on the experimental detection of this first light, Gamow turned Weber away. In 1965 this radiation was detected, almost by accident, by Arno Penzias and Robert Wilson, two Bell Telephone Laboratory scientists.

After Gamow's rejection, Weber decided to work on the physics of atoms with Keith Laidler, earning a Ph.D. in 1951 from The Catholic University of America. His thesis led to a paper that he submitted for presentation at an international conference in Canada on "coherent microwave emission." This paper had many of the key ideas and concepts that would lead to the maser (microwave amplification by stimulated emission of radiation) and ultimately to the laser (where "microwave" is replaced by "light"). Charles Townes was also working on this problem and asked Weber for a copy of his paper, while Nikolay Basov and Aleksandr Prokhorov in the Soviet Union were independently working along the same lines. In 1964, Townes, Basov and Prokhorov were awarded the Nobel Prize in Physics for their construction of the first masers and lasers. Although Weber was also nominated for the Nobel Prize at the same time, he never received it.

According to Kip Thorne, Weber's interest in general relativity began to grow after the maser discovery, because Weber wished to move into a field of study with less Nobel drama. In 1955 he took a sabbatical at the Institute for Advanced Studies to study gravitational radiation with Wheeler (their presentation at the 1957 Chapel Hill meeting was a result of that work), and then continued his studies at the Lorentz Institute for Theoretical Physics in the Netherlands.

And after these preliminary studies, he set out to do the unthinkable: to detect gravitational waves. Around 1958, he began the project in earnest, first doing the required theoretical calculations to determine just what the physical effects of a passing wave would be on a solid mass, in contrast to a set of disks sliding on a table, and then building an apparatus. By 1965 he had put a simple detector into operation. It consisted of a solid cylinder of aluminum (the reason for aluminum was a mundane one: it was cheap), about a meter in diameter by two meters in length. It weighed about 1.5 tons.

When a gravitational wave passes through the cylinder in a direction perpendicular to its axis, the spacetime distortion in the wave tries to

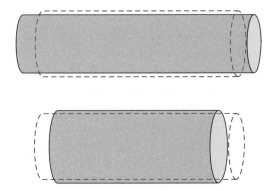

Fig 7.4 Distortions of a "Weber bar" induced by a gravitational wave traveling perpendicularly into the page that you are reading (vastly exaggerating the scale of the effect). Dashed lines indicate the undistorted cylinder. Top panel: The bar is stretched horizontally, while being squeezed in the other directions. Bottom panel: One half cycle later the bar is squeezed horizontally while being stretched in the other directions.

stretch and squeeze the bar end-to-end (see Figure 7.4). There is also stretching and squeezing in the perpendicular directions, but this turns out to be less easy to measure so we will ignore it. As we remarked before, because of the solid material in the bar between the ends, they do not move freely the way two sliding disks would with the same separation, and so the response to the wave can be tiny by comparison.

However, the bar has a property that the sliding disks do not. If you were to hit the bar on the end with a hammer, it would vibrate for a considerable time at a single frequency, called its "resonant frequency." Every child learns about the phenomenon of resonance at an early age. A playground swing moves back and forth with a characteristic frequency that depends on its length. If you push the swing with exactly the same frequency, achieved most easily by pushing once every cycle right when the swing returns to you (or if you are alone, by kicking your legs once each cycle at an optimal time), you can achieve thrilling amplitudes of swing. What's more, you continue to swing for a while even after the pushing has stopped, until friction from air or from the ropes brings the swing to a halt.

Weber had very good reasons for choosing a resonant bar as opposed to sliding disks. As a possible source of gravitational waves, he was thinking about supernovae, the only kind of "extreme" event known at the time. He imagined that the gravitational signal from such an event would be a short-lived "burst," possibly no longer than a small fraction of a second, and that it would be "broad-band," namely that it would contain waves with a broad range of frequencies as opposed to a single frequency, like a pure sound tone. If some part of the signal was at the resonant frequency of his bar, then the bar would be excited strongly, and in addition it would continue to oscillate at its resonant frequency after the burst had passed, giving more time for his sensors to measure the oscillations. The resonant frequency of his bar happened to be around a thousand cycles per second, or in the kilohertz band, in the same ball park as what one would estimate for the frequency of waves from a supernova. To measure the compression and extension of the bar, Weber bonded devices called "piezoelectric transducers" around the bar at the middle to convert the strains into electrical signals that could be recorded and analyzed. Nevertheless, it was still a daunting prospect: even the crudest estimate of the signal from a supernova in the Milky Way galaxy implied a change in length of his bar of about the diameter of a proton!

Weber devoted a great deal of effort to ensure that his bar was isolated from external disturbances, such as seismic vibrations or the effects of nearby traffic that could set the bar into oscillations and mimic the effect of gravitational waves. This required suspending the bars using the thinnest wires that would support the weight and attaching those wires to supports made of alternating layers of rubber and steel. He also shielded his bars from external electric and magnetic fields.

In June 1969, Weber made the stunning announcement that he was detecting signals simultaneously in two detectors spaced 1,000 kilometers apart, one in Maryland, and the other at Argonne National Accelerator Laboratory near Chicago. The reason for using two detectors is simply that any one detector is often in oscillation because of disturbances from the environment that leak in despite his best attempts to isolate the bar from such noise, and because of the inevitable random internal motions by the atoms inside the bar produced by heat energy.

Therefore, in a single bar it is difficult if not impossible to distinguish a disturbance from a gravitational wave from a disturbance of an environmental or thermal origin. As early as 1967, Weber had reported disturbances in a single bar, but he could not reliably claim that they were from gravitational waves. However, with two detectors separated by such a large distance, a disturbance that appears simultaneously in both detectors would be unlikely to be environmental or thermal because the probability of such a random coincidence is very small. Coincident events would therefore be good candidates for gravitational waves. Even more remarkable than the 1969 report of coincident events was his announcement in 1970 that the rate of such events was highest when the detectors were oriented perpendicular to the direction of the center of the galaxy, implying that the sources were indeed extraterrestrial, perhaps concentrated near the Galactic Center. These reports caused a sensation both in scientific circles and in the popular press.

There were two problems, however. The observed events occurred with a disturbance size and at a rate (around three times per day) that shocked theorists, for it implied a rate of gravitational wave bursts at least a thousand times what they predicted. This in itself was not necessarily bad, for often the mark of an important experimental discovery in physics is the degree to which it upsets theoretical sacred cows.

The second problem was more devastating, however. The main body of Weber's coincidence results were reported between 1969 and 1975. But by 1970, independent groups worldwide had built their own detectors with claimed sensitivities equal to or better than Weber's, yet between 1970 and 1975, none of these groups saw any unusual disturbances over and above the inevitable noise. By 1980 there was a general consensus that gravitational waves had not been detected by Weber. Weber never accepted any of this and he continued to work on the detection of gravitational waves with resonant bars until his death in 2000.

So, is this a story of a tragic failure or of a great success? It is certainly a good example of the self-correcting nature of science. The acceptance of new results always requires their external confirmation, typically done by carrying out the experiment again in a different setting and perhaps with more sophisticated instruments. In Weber's case, his results could not be replicated, so his claim was not accepted.

But as John Wheeler put it in 1998: "No one else had the courage to look for gravitational waves until Weber showed that it was within the realm of the possible."

Over time, a more nuanced view of Weber's legacy has emerged. Prior to Weber, the field of general relativity was almost completely dominated by theorists. The field was often called a "theorist's paradise and an experimentalist's purgatory." During the Chapel Hill conference, Feynman complained that the problem with general relativity is the lack of experiments. Weber's announcement induced experimentalists from other branches of physics to get involved in general relativity, such as William Fairbank from low-temperature physics, Ronald Drever from magnetic resonance work, Vladimir Braginsky from precision measurements, Heinz Billing from computer science, Edoardo Amaldi from elementary particle physics, J. Anthony Tyson from astronomy, and others. It also piqued the interest of a young MIT professor named Rainer Weiss, who would soon lay the foundations of the LIGO instrument. These experimentalists helped to transform the landscape of the field into one with a healthy synergy between theorists and experimentalists.

Weber's work also helped transform theory. As we have seen, when Weber started building his detectors, the main concern of theorists with regard to gravitational waves was their physical reality. After his announcement, the direction shifted dramatically as theory groups around the world starting thinking about plausible (and also implausible) astrophysical sources for the enormous signals he was claiming. Although no scenario was ever found that could explain Weber's signals, the insights gained and the techniques developed during this period helped to advance the growing interactions between general relativists and astrophysicists that had begun with the discoveries of quasars, pulsars and the cosmic background radiation during the 1960s.

Weber's work also had a personal impact on one of us. At the time of Weber's first announcement, Cliff had completed his first year as a Caltech graduate student and was thinking about doing a summer project in Kip Thorne's research group. Thorne told him:

I am worried that if Weber is correct, then general relativity itself might be wrong. I want you to spend the summer finding out all there is to know about the current experimental support for the theory and to think about what might be done in the future to prove the theory right or wrong.

That launched Cliff's fifty-year-long career in general relativity!

The consensus that Weber had not found waves did not end the effort to detect them, of course. Many groups continued to develop advanced bar detectors. One strategy involved cooling the entire bar and associated sensing devices to one or two degrees above absolute zero, in order to reduce the size of the disturbances due to the thermal motions of the atoms inside the bar. Some groups replaced the piezoelectric crystals used by Weber with sophisticated sensors attached to the ends of the bars, leading to greatly improved performance. Some fabricated their bars out of different materials such as sapphire that might have an improved response to the gravitational wave excitations.

None of these groups reported credible detections, but instead established increasingly stringent upper limits on the strength of gravitational waves bathing the Earth. Although by 1979, as we saw in Chapter 5, the measurements of the binary pulsar had verified the existence of gravitational waves, the actual waves emitted by that system were far too weak and of too low a frequency to be detectable by resonant bars. Work continued on bars for another 25 years, but gradually declined for lack of funding until the last "Weber bars" ceased operations for gravitational wave detection around 2008. But important advances were made in the course of this research, in new techniques for isolating the bars from things like seismic noise, in the control of thermal noise and in data analysis techniques. Many of these lessons would be used in helping to develop an alternative detector concept that began to emerge during the 1970s, the laser interferometer.

The laser interferometer is based on an apparatus devised by US scientist Albert A. Michelson originally to measure the speed of light very accurately, but then famously used by him and Edward W. Morley

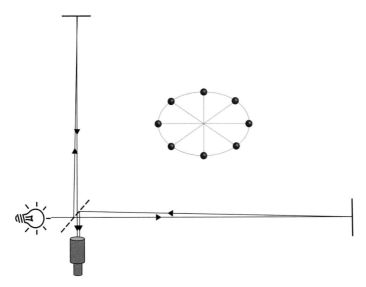

Fig 7.5 Schematic version of a laser interferometer. Light from a laser is split by a half-reflecting mirror and travels along two perpendicular arms. Mirrors at each end reflect each beam back. The beams are brought together at a sensor. If the waves interfere constructively, a bright spot is seen; if they interfere destructively, a dark spot is seen. A "plus"-polarized gravitational wave that travels perpendicularly into the page you are reading will stretch one arm while compressing the other, thus altering the interference between the two light beams.

in 1887 to try to detect the motion of the Earth through the "aether," the hypothetical medium through which light supposedly propagated. Schematically, Michelson's interferometer consists of two straight arms set at right angles to each other (Figure 7.5). Each arm has a mirror at one end. At the intersection where the arms are joined, a half-silvered mirror splits a light beam into two, each traveling down one arm, each reflected back by the mirror at the end of each arm. When the two beams recombine, they interfere to produce a characteristic pattern of fringes that depends on the difference in time required for the two beams to make the round trip. Michelson and Morley failed to detect any effect of motion through the aether, and the conundrum inspired by that failure ultimately led to Einstein's special theory of relativity.

Michelson's interferometer is an excellent tool for measuring distance precisely, because a change in the length of an arm of a quarter of the wavelength of light will turn a bright spot at the output into a dark spot. Since the wavelength of light is measured in millionths of a meter, it is easy to see the potential, at least in principle, of measuring the tiny changes in separation between objects induced by a gravitational wave. The reader might again ask: the gravitational wave changes the distance between the mirrors and the beam splitter, but doesn't it also affect the propagation of the light ray? The answer is yes, but just as in the example of the disks sliding on the table, the two effects do not cancel each other out, and there is a true, measurable change in the brightness of the output beam.

The first to think about such a scheme were the Russian physicists M. Gerstenshtein and V. Pustovoit in 1963, but their work was not recognized until many years later. Weber and his student Robert Forward independently considered this possibility, and Forward actually built the first prototype for an interferometric detector in 1972. But the person generally credited with showing how to turn this idealized concept into a large-scale device that might actually detect gravitational waves is Rainer Weiss, or Rai, as he is known to all his colleagues.

Rai's family escaped Nazi Germany and relocated to New York City in 1939. A brilliant tinkerer in anything electronic, he started his studies in electrical engineering at MIT, but left Boston in his junior year to pursue a romantic relationship in Chicago that would eventually fizzle out. Upon his return to MIT, he discovered he had been expelled because he had gone AWOL. Undaunted, he managed to convince MIT physicist Jerrold Zacharias to give him a job as a technician in his lab. Around that same time, Zacharias was working on the first practical version of an atomic clock, based on cesium atoms. With the help of Zacharias, Rai was readmitted to MIT, finished his undergraduate degree in 1955 and finally obtained a Ph.D. under Zacharias in 1962.

After a two-year stint as a postdoc with physicist Robert Dicke at Princeton University, where he began to develop experiments to test general relativity, he returned to MIT in 1964 as an assistant professor. At the urging of radio astronomer Bernard Burke, he took an interest in the cosmic microwave background radiation, recently detected by Penzias and Wilson. This radiation is the remains of the hot electro-magnetic radiation that would have dominated the universe in its earlier phase, now cooled to a few degrees above absolute zero by the subse-quent expansion of the universe. Cosmological theory suggested that the strength of this radiation should have a very specific dependence on the wavelength of the radiation, known as a "black-body spectrum," but measurements made using sensors on rockets threatened to discredit this prediction. Rai and his graduate student flew a device on a high-altitude weather balloon that showed convincingly in 1973 that the spectrum was that of a black body, and also measured its temperature to be 3 degrees above absolute zero (the modern measured value is 2.725°). Rai would later be a leader on NASA's Cosmic Background Explorer satellite, which would make even more precise measurements of the properties of this radiation between 1989 and 1993.

But he had never truly forgotten about his first love, gravity experi-ments. Around 1968, MIT asked him to teach a class on general relativity, but not being an expert on the theoretical side of the subject, he chose to focus on the experimental side, using a small 1961 primer on general relativity and gravitational waves that had been written by Weber. But when he studied Weber's discussion of using resonant bars, he could not understand how Weber could achieve the sensitivity needed to detect the waves. So he assigned a homework problem in his course: find a way to measure gravitational waves by sending light beams between things. The students were stumped, so not much came out of that homework set. Soon Weber was claiming detections, to Rai's skepticism, and so he decided to think more seriously about the interferometer idea. He analyzed in detail every source of noise or disturbance he could think of and concluded that, with a large enough interferometer, it might be possible to beat Weber's sensitivity by a factor of a thousand. He also recognized a key difference between Weber's resonant bars and an interferometer. The bars respond strongly only to the part of the gravitational wave whose frequency is close to the resonant frequency of the bar, whereas if one suspended the mirrors in the interferometer on long pendula, they would respond fully to the gravitational wave, no matter what its frequency was, just like our frictionless hockey pucks.

Thinking that this analysis was not worth publishing as a scientific paper, he wrote it up in a twenty-three page report printed in one of MIT's quarterly newsletters in 1972. The concepts laid out in his report would become the foundation of LIGO's design.

While preparing his MIT report, he requested and obtained funding from MIT to build a small prototype with arms 1.5 meters long, and around 1975 he wrote a proposal to the NSF to continue this work. Despite positive reviews, the proposal was turned down. The proposal came to the attention of Heinz Billing at the Max Planck Institute in Munich. One of the pioneers of computer science, Billing had recently returned to physics, and his group was engaged in using resonant bars to check Weber's claims. Turned on by Rai's description of the potential of interferometry for detecting waves, he and his colleagues started to build a prototype. Soon his laboratory was paid a visit by Ron Drever, who also had been working on bar detectors, and now became intrigued by this new approach. Drever (1931–2017) was a brilliant and inventive physicist at the University of Glasgow, who at the age of twenty-nine performed an exquisite experiment using nuclear magnetic resonance techniques to show that the mass of an atom does not depend on its orientation relative to the Galaxy or relative to the Earth's velocity through the universe. Drever's group began to build a prototype interferometric detector.

Kip Thorne at Caltech also began to think about interferometers, spurred by a late-into-the-night discussion with Rai Weiss at a hotel in Washington during a NASA committee meeting. His group had been at the forefront of the theory of gravitational wave sources, but he felt that Caltech should also have a presence in the experimental side, and so he recruited Drever to move to Caltech in 1979, where he began to build a 40 meter prototype.

It had become clear that tabletop or laboratory-scale prototype interferometers were fine for technology development, but that they would not be sufficiently sensitive to ever detect the kinds of gravitational waves that might reasonably be expected from astrophysical sources.

Instead, devices with arms as long as many kilometers would be needed. The reason is that the change in distance between two objects that a gravitational wave induces is proportional to the distance. If you double the distance between the beam splitter and the mirror, you double the displacement, and therefore you double the difference between the light beams when they recombine. Go from a 40 meter prototype to a 4 kilometer system and you increase the effect by 100. Unfortunately, you also increase the cost by a similar factor. One reason is that the light must propagate through an ultra-high vacuum, otherwise the fluctuations in the light speed caused by its interactions with the atoms in the residual gas or air would have a larger effect than the displacements of the mirrors caused by a gravitational wave. There was also general agreement that, just as with Weber's bars, two widely separated interferometers would be needed in order to claim a credible detection. It was thus also becoming clear that gravitational wave detection would be very costly.

As a result, at the urging of the NSF, Caltech and MIT agreed in 1984 to cooperate on the design and construction of LIGO, with joint leadership by Weiss, Thorne and Drever. This leadership arrangement proved unworkable, however, and in 1987, astrophysicist and former Caltech Provost Rochus E. Vogt was appointed LIGO director. By 1992, initial funding for construction had been provided by the US Congress, and sites in Hanford, Washington and Livingston, Louisiana had been selected. Barry Barish, a high-energy particle physicist, replaced Vogt as LIGO director in 1994 and oversaw the construction and commissioning of the detectors and the initial gravitational wave searches. The plan for LIGO involved two stages: building and operating the interferometers with proven technology, with a sensitivity where gravitational waves *might* be detected, and then to upgrade them with advanced technology to a level where waves would be detected, if general relativity and our understanding of astrophysics were correct. Searches for gravitational waves with the initial LIGO were carried out between 2002 and 2010. To nobody's surprise, no waves were detected. On the other hand, the interferometers reached the sensitivities they were designed to achieve and much experience was gained in operations and data analysis. Between 2010 and 2014, the interferometers were shut down in order to install advanced technology that had been under development, such as more powerful lasers, improved mirrors and better isolation from seismic disturbances. By September 2015, the instruments were as much as ten times more sensitive than in the initial LIGO.

But the Americans were not the only ones who wanted to detect gravitational waves. Alain Brillet was a French physicist at the National Research Center in Orsay near Paris who had worked with Jon Hall at the University of Colorado in 1979 to do a twentieth-century version of the famous Michelson–Morley experiment, but using a laser as the light source in the interferometer. Adalberto Giazotto was an elementary particle physicist at the National Institute of Nuclear Physics in Pisa, Italy who had taken an interest in gravitational wave detection, and particularly in the problem of seismic isolation of the mirrors. Together they proposed a large European interferometer, which was ultimately built near the town of Cascina, about 15 kilometers south-east of Pisa, and named Virgo after the Virgo cluster of galaxies. Drever's group in Glasgow and Billing's group in Munich combined forces to propose a large interferometer. Because of funding limitations, caused in part by the cost of the reunification of Germany, they had to settle for 600 meter arms, compared to the 4 kilometers of LIGO and 3 kilometers for Virgo. That instrument, called GEO-600, was built near Hannover, Germany. Researchers in Australia initially built advanced resonant bar detectors, and then moved into interferometers, but could never convince their government to go beyond an 80 meter prototype called AIGO, sited near Perth in Western Australia. Japanese teams also became very active, and have recently completed an ambitious interferometer, the Kamioka Gravitational Wave Detector (KAGRA), a 3 kilometer instrument built deep inside Mount Ikeno near Hida, Japan, where numerous underground physics experiments studying neutrinos, dark matter and proton decay have been run using inactive shafts and tunnels from the Kamioka mine (see Chapter 9).

You might be picturing an intense international race and competition to be the first to detect gravitational waves, but in fact quite the opposite happened. Recall Weber's dictum that you had to have more than one detector to be sure that you have detected gravitational waves. In addition, since the mirrors in the interferometers respond immediately and freely to a passing gravitational wave, by recording the difference in time of arrival of the same signal in two widely separated interferometers, you can get some idea of the direction of the source. The principle is the same as the one we discussed in Chapter 3 (see Figure 3.5), whereby the difference in arrival time of a radio wave at two separated radio telescopes can be used to determine the source direction. Two interferometers give only limited information about the location of the source on the sky; the more interferometers you have, the more accurately you can pinpoint the source. For this to work, different teams have to cooperate, regardless of the desire of any individual or national agency to garner the glory of being "first." So while LIGO and Virgo were still under construction, the leadership of the two projects began the delicate negotiations that would ultimately lead in 2007 to the LIGO–Virgo Collaboration, a rather remarkable structure that views the two LIGO instruments and Virgo as a single network of three interferometers, with full data sharing and transparency, coordination of schedules, and so on. (For colleagues of ours who are members of the collaboration, it also means an unbelievable number of teleconferences across many continents, resulting often in highly inconvenient work hours!) The GEO-600 and AIGO teams joined the collaboration to work on technology development. Despite its lower sensitivity, GEO-600 also made observations when the LIGO and Virgo instruments were offline, just in case a strong event, such as a nearby supernova, might occur. Virgo also adopted a two-stage development strategy, similar to the initial-to-advanced LIGO track. When the first detection was made on 14 September 2015, advanced-Virgo was still about a year away from being up and running, so the signal was only seen by the LIGO interferometers, yet the discovery paper published in 2016 included all the members of Virgo as co-authors. The paper had over a thousand authors. In the next chapter we will see how important this cooperation proved to be, when we discuss what signals were actually detected and what they implied.

By early 2015 both LIGO interferometers were working, and entered what is called an "engineering run," during which the operators of the instruments poke and prod them, tweak the dials and alter various settings, all in an effort to get the maximum performance. That run was scheduled to end on 18 September, when all tweaking would cease and an "observing run" would begin. But the engineers completed their work about a week ahead of schedule, and both interferometers were performing quietly, awaiting official kick-off of the observing run. On 14 September at 5:51 a.m., Eastern Daylight Time, the Livingston instrument recorded a signal, and 7 milliseconds later the Hanford detector recorded the same signal. The signal, known thereafter by the name GW150914 (GW for gravitational wave, followed by the date in yymmdd format), arrived close to a hundred years after Einstein published his theory. In the next chapter we will describe the detection and what we learned from it.

Before we do that, there is one final aspect of gravitational waves that we need to discuss. In various newspaper reports about gravitational waves, you may have encountered the phrase "listening to gravitational waves." Many popular books on this subject use musical motifs, such as Marcia Bartusiak's *Einstein's Unfinished Symphony* or Janna Levin's *Black Hole Blues*; Chapter 9 of this book talks about a "loud" future instead of a bright future. What is that all about? Normally you think about astronomers "gazing" at the heavens, "seeing" a supernova explosion, or "watching" a planet transit in front of the Sun. Why do we "listen" to the universe with gravitational waves?

The reason has to do with the fundamental difference between electromagnetic waves and gravitational waves. When electromagnetic waves, a.k.a. light, impinge upon some material, such as the retina of your eye, the electric and magnetic fields of the light waves push on the charged electrons in the material and generate an electrical current. If you prefer the quantum mechanical picture in which light consists of "photons," then the photons knock the electrons loose from their host atoms. The current is then sent from the retina of your eye to the optic nerve and then to your brain. The current could be produced in the CCD device in your camera or smartphone. Or it could be in the conducting antenna of a radio telescope. The act of "seeing" in all its manifestations basically amounts to using light to move electrons and thereby to produce electrical currents.

Gravitational waves act very differently, causing bits of mass (not charge) to move back and forth relative to each other via the stretching and compressing of spacetime (recall the hockey pucks in Figure 7.3). So when a gravitational wave passes through your head, it causes the eardrum and bones in one ear to move relative to those in the other ear. It also tries to stretch and compress your skull in the same manner, but since your skull is rather rigid, it can resist that effect to a large degree. The elements of your inner ear, on the other hand, are more free to move, and in doing so, they strike the membrane of the cochlea, forcing the fluid inside to move back and forth, triggering a set of hairs that convert the oscillations into electrical impulses that travel through nerves to your brain. The only difference between sound waves and gravitational waves in this regard is that sound waves use the expansion and compression of air to move the eardrum, while gravitational waves use the expansion and contraction of spacetime itself to produce the same effect.

But we cannot sense all possible vibrations of our eardrums. This is because the conversion of the oscillations into electrical impulses is not efficient enough for frequencies below about 20 hertz (for the lowest-pitched sounds) or above about 20,000 hertz (for the highest-pitched sounds). Other mammals, such as dogs, can hear up to 45,000 hertz, while the greater wax moth can hear sounds up to 300,000 hertz! Dog whistles, in fact, leverage this physics principle: they produce vibrations in the air that are too high a frequency for the human ear to pick up, but that can be easily heard by dogs.

As we will learn in the next chapter, the waves from GW150914 had a frequency in the region between 40 and 300 hertz, well within the human audible band. So why didn't people around the world hear the signal? The answer is the incredible weakness of gravitational waves. The minimum eardrum motion that we can detect is about a nanometer, or a millionth of a millimeter. The gravitational waves detected by LIGO moved our eardrums by about a trillionth of a nanometer! Alternatively, we could have heard those waves if the source had been extremely close, about twice as far from the Sun as Neptune, instead of over a billion light years away. We should be happy that this was not the case. Since the source was two black holes each about thirty times the mass of the Sun in a close orbit around each other, their normal gravitational tug on the Sun and planets would have disrupted the Solar System (or prevented its formation in the first place) long before we came along.

Instead of eardrums, we have the mirrors of LIGO and Virgo. Instead of the tiny bones of the inner ear and the cochlea, we have laser beams bouncing off the mirrors, capable of detecting their motions to better than a thousandth of the diameter of a proton. The laser interferometers of LIGO and Virgo are our tools for listening to gravitational waves. And what have we learned from these sounds?

CHAPTER 8

What Do Gravitational Waves Tell Us?

T minus 2 hours: At the LIGO Hanford observatory in Washington State it is 12:50 a.m., and the last two scientists depart to get a well-deserved night's sleep after a long day of engineering tests, leaving only the night-duty operator on site. The detectors had been turned on a few months before to calibrate the instruments in preparation for the first science run of the "Advanced LIGO" detectors, which is scheduled to begin in four days.

Two billion kilometers away, about one and a half times the distance to Saturn, a packet of spacetime ripples approaches the solar system at the speed of light (Figure 8.1).

T minus 15 minutes: In Livingston, Louisiana it is 4:35 a.m. and two technicians at the LIGO Livingston observatory are about to run one last test before finishing their shift. The "bump test" consists of driving a car at 5 or 10 miles per hour over speed bumps right outside the LIGO building, GPS unit in hand, to check whether the motion of the car over the speed bumps creates ground vibrations strong enough to be detected by the interferometer. Right before starting the test, they realize their GPS unit is malfunctioning and needs to be recalibrated, so instead of running the test they call it a night and go home.

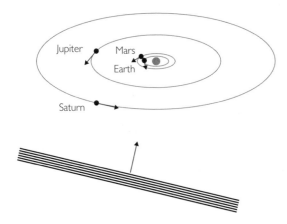

Fig 8.1 A packet of gravitational waves approaches the solar system on 14 September 2015. The waves come from below, about 15° from the "south pole" of the solar system. Shown are the positions of the Earth, Mars, Jupiter and Saturn at that time. The waves from this source may have been passing through the solar system for millions of years; we only show the final burst of strong waves, the ones detected by the LIGO instruments.

The packet of spacetime ripples is now 270 million kilometers, or 1.5 astronomical units, away.

T minus 0 minutes: The packet of spacetime ripples passes through the LIGO Livingston detector, and 7 milliseconds later it passes through the LIGO Hanford detector. For about a fifth of a second the mirrors at opposite ends of the interferometers move back and forth a few times by a tiny amount and the interference patterns of the laser beams inside the instruments change a bit, generating electrical signals that record the passage of the spacetime ripples. The signals are recorded automatically, but nobody notices. Not just yet.

T plus 10 minutes: It is noon in Hannover, Germany, close to lunch time. Postdoctoral researcher Marco Drago at the Albert Einstein Institute notices that an automated computer program has pinged, announcing the presence of something strange in the LIGO data. He is curious and so he checks the logs to see if there was a scheduled injection, a test to make the mirrors in the instrument vibrate artificially just as if a wave had gone through. But the logs show no scheduled injection. Marco goes next door to the office of his friend, Andrew Lundgren, also a postdoctoral researcher. He tells Andrew about the ping and together they investigate it further. Could it have been an injection that was not logged? No, there was truly nothing scheduled. Could it have been a bump test? No, there was nobody testing the interferometers. Could it have been a micro-earthquake or some atmospheric effect? No, the data quality monitors were all showing perfect conditions.

T plus 54 minutes: Marco sends an email to the entire LIGO scientific collaboration, over 1,000 researchers spread around the world. He describes the event that was recorded and he ends the email asking for confirmation that this was not an artificial injection. A flurry of emails ensues. Within hours, there is confirmation from the LIGO leadership that this event was *not* an injection or test of any kind.

T plus 10 hours: The LIGO executive committee gathers via telephone conference call. They discuss the event and make a decision: maintain the instruments in their current state and continue taking data. The software is locked. The hardware is locked. Cabinets housing the electronics are physically locked. For the next two weeks, nobody is to touch either of the two instruments or alter a setting, so that more data can continue to accumulate to be able to compare the event to data that presumably contains only random noise.

The event prompts an "all hands on deck" response by the thousand-member team to verify or refute the idea that this was a gravitational wave detection. Several independent computer analyses of the data reveal the same signal in each detector. One simple analysis applies a technique that is similar to that used in high-end headphones and hearing aids to cancel part of the background noise so that you can hear the music or dialogue you are interested in, even in a noisy airplane or restaurant. Known as a "band-pass filter," it suppresses noise in the frequency range above and below the frequency range where the signal

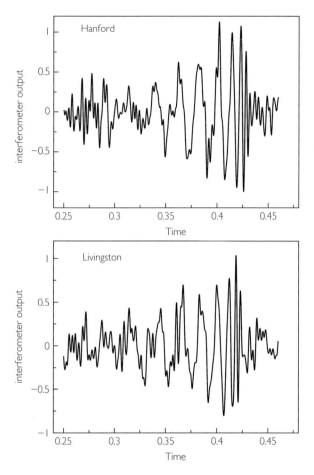

Fig 8.2 Data from the Hanford and Livingston detectors during the crucial 0.2 seconds, after being run through a band-pass filter, and after certain noise effects associated with well-understood vibrations in the instrument have been removed. Credit: Gravitational Wave Open Science Center.

resides, between about thirty and a few hundred hertz, while not altering anything in that critical frequency range. What emerges when that simple filter is applied to the data from each detector is shown in the two panels of Figure 8.2.

Each panel shows a stretch of filtered output lasting about two tenths of a second, occurring at the same time in each detector. The clocks at the detectors are all set to Greenwich Mean Time in order to avoid any possible confusion with time zones or daylight savings time. From about 0.25 seconds to about 0.34 seconds the outputs look very spiky and do not resemble each other at all. This is the random, independent noise present in each detector. From 0.34 seconds to about 0.38 seconds we see three peaks and valleys that roughly match each other in each detector, but with some spiky noise superimposed. Those three peaks represent two complete cycles of the wave, over a time of 0.04 seconds, corresponding to about 50 cycles per second or 50 hertz. These are followed by four more peaks and valleys that are significantly higher than the first three, but also are more closely spaced than the previous peaks. Those three full cycles span only about 0.025 seconds, corresponding to a frequency of about 120 hertz. This implies that not only is the strength of the signal increasing with time, but its frequency is also increasing with time. But the last of these four peaks in each detector is already lower than its predecessor. After this, the output looks once again like independent random noise in each detector.

Even if you had no idea what this signal represented, you would be tempted to think that this was a candidate gravitational wave signal. First, the signal is almost exactly the same in the two detectors. To be sure, it is conceivable that some event, such as a tree falling near the Livingston detector (the surrounding forest there happens to undergo active logging) could, by some fluke, produce exactly the right vibrations of the ground to produce the signal seen in the bottom panel of the figure. But those ground vibrations in Louisiana could not possibly affect the Hanford detector, 3,000 kilometers away in Washington State. And the chance of some unrelated event at Hanford (which is surrounded by almost treeless high desert) producing exactly the same response at exactly the same time is astronomically small (we will quote a number later). This is one of the positive legacies of Joseph Weber's failed attempt to detect gravitational waves, the principle that for a claimed detection of gravitational waves to be credible, the same signal *must* be sensed in independent, widely separated detectors. The fact that the two signals are not *exactly* the same reflects the everyday fact that two people listening to a third person in a very noisy room might not hear exactly the same thing, but will still get the gist of what is being said.

Another feature of the two signals is important. While the peaks and valleys in the two panels seem to line up in time, the features in the Hanford detector are consistently about 7 milliseconds (0.007 seconds) later than those in the Livingston detector (the difference is too small to show up on the figures, but it is easily measured from the data). Now, if the gravitational waves were propagating from somewhere in the sky exactly perpendicular to the line joining the Livingston and Hanford detectors (the baseline), they would arrive at the two detectors at exactly the same time (see Figure 8.3). If they were traveling exactly parallel to the baseline, then they would arrive at one detector 10 milliseconds before the other. This is the time it would take a signal traveling at the speed of light to traverse the 3,000 kilometers between the two detectors. The actual time difference of 7 milliseconds was comfortably between these two limits, indicating that the signal actually arrived from a direction about 45 degrees from the baseline (right panel of Figure 8.3). On the other hand, if the time difference had been greater

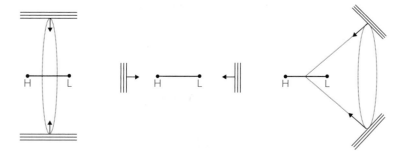

Fig 8.3 Left: Waves approaching the Hanford and Livingston detectors from any direction perpendicular to the line joining them will reach the detectors at the same time. Middle: Waves approaching parallel to the line joining the detectors will reach one detector 10 milliseconds before the other, because of the 3,000 kilometers separating them. Right: Waves approaching from a direction approximately 45 degrees relative to the baseline will reach the Livingston detector about 7 milliseconds before the Hanford detector. The measured time differences give important information about the location of the source on the sky.

that 10 milliseconds, this would *not* have been accepted as a candidate gravitational wave.

The researchers at the LIGO scientific collaboration actually had a very good idea of what the signals in Figure 8.2 represented: the "chirp" signal from the final inward spiral and merger of two bodies such as black holes or neutron stars. We will describe the history and physics of this idea later in this chapter, but the bottom line is this: as the two bodies orbit each other they emit gravitational waves, thus losing energy, getting closer to each other and orbiting faster (recall the binary pulsars from Chapter 5). This "inspiral" phase leads to waves of increasing strength or amplitude and increasing frequency, as shown in Figure 8.4. This part of the signal is called a chirp because of the similarity between a sound with these characteristics and the songs of some birds. The two bodies then merge, forming a black hole, a process that leads to a brief burst of strong waves (the "merger" part of Figure 8.4). The newly formed black hole is very distorted and it oscillates or "rings" a few times, emitting "ringdown" waves and quickly settling down to a stationary black hole that ceases to emit gravitational waves. The wave shown in Figure 8.4, calculated using an approximate solution from general relativity, displays all the features shown in the two panels of Figure 8.2.

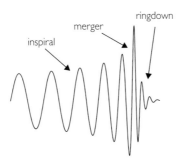

Fig 8.4 A chirp signal from two black holes calculated using general relativity, showing the inspiral part, when the two black holes are orbiting each other with increasing speed, the merger part, when the two holes merge to form one very distorted black hole, and the ringdown part, when the distorted black hole emits gravitational waves and settles down to a final stationary black hole.

One can imagine the level of euphoria that occurred within the collaboration. But with this euphoria came an accompanying sense of paranoia. What if somebody maliciously inserted an artificial signal just to fool us? Surely this is unlikely. Hackers are not typically interested in astronomical data sets, and it would be unfathomable for a collaboration member to be so malicious.

More worrisome was the possibility that some bizarre noise artifact, erroneous instrumental setting or faulty line of computer code was making them believe they had detected a gravitational wave, when in reality they had not. It would not be the first time in the history of physics that an erroneous claim had been made. After all, physics is done by people, and people make mistakes. The important fact about physics is that it is self-correcting, and errors are eventually fixed and the record is set straight. But nobody wants to be known for the mistake rather than the discovery.

Luckily, examples of such errors are rare in physics. But when they occur, they generally make headlines and cause much embarrassment.

In 1989, electrochemists Stanley Pons and Martin Fleischmann announced they had observed "cold fusion" in their lab. Nuclear fusion occurs regularly inside the Sun, converting hydrogen into helium, releasing energy sufficient to warm and illuminate the Earth (the same process occurs in thermonuclear bombs). But this process requires extremely high temperatures. Achieving fusion at room temperature would have been revolutionary, as it would provide an effectively limitless energy source. Many scientists tried immediately to replicate their experiment but most failed, and in time the flaws in the original experiment that had led Pons and Fleischmann to the wrong conclusion were identified. Apart from the effect on the careers of the two scientists, the episode was a major embarrassment for the University of Utah, which had exploited the discovery for maximum publicity.

In 2011, an experiment with the acronym OPERA made a dramatic announcement. The instrument was designed to study subatomic particles called neutrinos emitted in the CERN accelerator in Geneva, Switzerland and directed toward detectors 730 kilometers away, inside the Gran Sasso mountain in Italy. In September of that year, the OPERA collaboration announced they had measured an anomaly that might be a sign of neutrinos traveling faster than the speed of light. Such a discovery, if correct, would have been revolutionary, as it would have contradicted Einstein's special theory of relativity, and consequently, his general theory. But a few months later, the OPERA team reported two flaws in their equipment: one related to a fiber optic cable that was not connected properly and another related to a clock that ran fast. These flaws, they concluded, were responsible for the anomaly. After correcting the problems, they found that neutrinos indeed travel at the speed of light, up to measurement uncertainty. But in the end, OPERA is remembered more for the mistake than for the final valid result.

And finally, there is the internal history of gravitational wave science. As we described in Chapter 7, an announcement of the discovery of gravitational waves had already been made in the late 1960s by Weber. Immediately after this announcement experimentalists set out to replicate Weber's results, but they failed. Eventually, a consensus arose that Weber's result had to be wrong, so when the LIGO detectors were built, the collaboration wanted to be particularly careful to not make the same mistake again. Many checks and counter-checks were established and tested to ensure that a detection was real prior to any announcement of a discovery.

This system of checks was so rigorous that it led to what is now known as the infamous "Big Dog" event. On 16 September 2010, an initial, less sensitive version of the LIGO detectors was in science mode, collecting data in the (admittedly unlikely) event that a sufficiently loud gravitational wave would pass through the Earth. And on that day, the alarms went off. A candidate event was identified, and it seemed to be coming from somewhere near the direction of Sirius, the Dog Star. In a fit of creativity, the event was named the "Big Dog." Eight minutes after the event was detected, roughly twenty-five people in the collaboration were notified to follow up on it and see whether it was worthy of further study. These twenty-five people concluded that this was the case, and the collaboration sent a circular to a group of collaborating astronomers to tell them that a candidate event had been detected, while they continued to analyze the data.

Everybody involved was sworn to secrecy because the Big Dog could have been a false alarm, either from a rare simultaneous disturbance at both detectors, or from a "blind injection." A blind injection is an internal test carried out routinely by the collaboration in which a tiny group of pre-selected technical people in the collaboration inject a fake signal in the data without telling anybody else, except an even tinier group of pre-selected VIPs in the collaboration. The purpose of this test is to see if the automated data analysis tools they have created can catch the blind injection and if the collaboration can identify it properly. For several months, the collaboration carried out all the tests and checks on the Big Dog, verified that the signal represented a gravitational wave, and wrote a draft of the discovery paper. On 14 March 2011, the "envelope was opened" and (drum roll) the LIGO leadership announced to the team that the Big Dog had been a blind injection after all. The good news was that the collaboration had caught it and so the data analysis tools were working as expected. Well, not exactly: some of the inferred parameters, like the location of the source in the sky, were not the same as those of the injected signal. This led to the discovery and correction of a line of computer code with a wrong sign. The bad news was that they hadn't detected a real signal.

You might now understand why, in September 2015, when the data analysis tools signaled that a candidate event had just been detected, the collaboration was extremely cautious and secretive. Not satisfied with the simple filter that revealed the signals shown in Figure 8.2, they analyzed the data using sophisticated computer code and managed to extract a beautiful chirp using a sum of short waves of fixed frequency. This was an important test because it was agnostic to the true theory of gravity, using almost no information about Einstein's theory. Simultaneously, the collaboration also compared the data to an array of detailed general relativity predictions of the gravitational wave signal emitted by merging black holes, finding agreement with their initial conclusions based on cruder analyses. Those comparisons also allowed them to measure such quantities as the masses of the two black holes, and to test Einstein's theory. We will describe what was learned in a moment.

The collaboration also calculated the probability that this was a fluke, a pair of random events at each detector that happened to jiggle the mirrors just so. After ten million computer simulations, they found that such an accident would happen less often than once every 200,000 years. So this was not a fluke, and it was not an injection of any kind. In fact, right after the detection, LIGO management had revealed that there was no "Big Dog"-style blind injection. This event was the real thing.

This extraordinary degree of caution, secrecy and obsessively detailed analysis explains why it took five months from the initial "ping" that caught Marco Drago's eye to David Reitze's announcement at the National Press Club in February 2016.

As we have said, the first detection was of waves from the final few inspiraling orbits and the merger of two black holes. This seems like a ridiculously special, once-in-a-lifetime event. Although we knew that binary neutron stars exist (see Chapter 5), there was no observational evidence that binary black holes exist. Surely a much more plausible possibility for the first detection would have been a supernova, examples of which had been observed by humans for millenia. These were the sources that Joe Weber was after when he built his resonant bar detectors.

But in fact, theorists had been thinking about black hole or neutron star inspirals and mergers for some time, and by the time the LIGO detectors were being considered by the NSF for major funding, gravitational waves from inspirals and mergers already formed the centerpiece of the science case that LIGO advocates were making.

Strangely enough, the idea was first proposed in 1963 by physicist Freeman Dyson while he was studying how advanced extraterrestrial civilizations could sustain their energy needs. Born in England in 1923, Dyson moved to the US in 1947 to study for a Ph.D. at Cornell (although he never actually received the degree). Over his seventy-year career (he is a professor emeritus at the Institute for Advanced Studies in Princeton) he made important contributions to an eclectic array of scientific subjects, including pure mathematics, quantum field theory (in 1947 he proved that the seemingly discordant theories for quantum electrodynamics that had been devised by Richard Feynman and by Julian Schwinger were actually different versions of the same theory, today called QED), biology and space exploration, as well as topics in the public interest, such as nuclear warfare and climate change. In 1955 he met Joe Weber during Weber's sabbatical with John Wheeler, and became intrigued with the idea of detecting gravitational radiation.

However, in 1963 he was interested in whether there was a better source of energy to sustain an advanced extraterrestrial civilization than the light and heat from its host star. In an article entitled "Gravitational Machines" he imagined a civilization stationing its home planet or base station not too far from a binary star system. If the civilization sends a probe toward one of the stars, allowing it to make a close flyby of the star at a time when the star is approaching the base station, then the probe would return to the base station with more kinetic energy than it had when it departed. That energy could then be extracted and used to sustain the civilization. The effect he was employing in his model is called the "gravitational slingshot," well known to planetary scientists as a way to boost the speed of spacecraft to higher levels than they could ever acquire from rockets; it is routinely exploited to get spacecraft to Jupiter and Saturn and beyond, for example. The problem with Dyson's idea is that binary stars typically move so slowly in their orbits that the energy obtained from the slinghot effect is trivial compared to the conventional energy from the light of the stars. A binary system of white dwarfs would be better because, being a hundred to a thousand times smaller than a solar-type star, they can orbit more closely to each other and thus achieve much higher velocities. This would be more promising for the civilization, particularly since white dwarfs are too dim to provide sufficient light and heat.

But then, Dyson reasoned, even better would be a binary system of neutron stars. These bodies are so small compared to their masses—roughly 20 kilometers in diameter—that they can orbit each other in very close proximity and with speeds that are a significant fraction of the speed of light, and thus the energy available to the civilization on each slingshot is even larger. This was quite a radical idea in 1963, since, as we saw in Chapter 5, neutron stars were at the time little more than a figment of Baade's and Zwicky's imaginations, and the first neutron star, in the form of a pulsar, would not be detected until four years later.

Of course, there was also no evidence of extra-solar planets at the time, let alone other civilizations. In any event, Dyson immediately realized that this idea would not work. Such a close neutron star binary would emit copious amounts of gravitational radiation, and would then inspiral and merge so quickly that the civilization would soon lose its source of energy. On the other hand, he noted, the gravitational wave signal itself might be of interest, prophesying that "[it] would seem worthwhile to maintain a watch for events of this kind, using Weber's equipment or some suitable modification of it."

But because Dyson's paper appeared in a book on extraterrestrial life, nobody working in general relativity noticed it or followed up on it. It was the discovery of the binary pulsar in 1974 that made the idea of binary inspiral respectable (see Chapter 5). The binary pulsar was proof that binary neutron stars exist, and the measurement of its decreasing orbital period proved that such systems spiral inward because of the emission of gravitational waves. As Taylor and his collaborators found, the rate of inspiral is absolutely tiny, only 76 microseconds per year in its orbital period, or 4 meters per year in separation. But the measured rate agreed with the prediction of general relativity. The theory goes on to predict that as the orbit shrinks, the stars speed up, thereby emitting stronger gravitational waves, which thus accelerates the inspiral, leading to even stronger waves, and so on, culminating in a runaway rush toward a final merger. For the Hulse–Taylor system, the formula predicted that the merger would occur in a few hundred million years, while for the double pulsar, it would occur in 85 million years. These are ridiculously long times—don't bother to mark your calendars—but they are only a few percent of the thirteen-billion-year age of the Milky Way galaxy. So one could easily imagine counterparts to these binary pulsars starting their evolutions a few hundred million years ago and merging today, thus fulfilling Dyson's prophesy.

As planning for large-scale laser interferometers moved forward, the idea that neutron star inspirals might be a leading potential source of gravitational waves began to take hold. The discovery of a few more binary pulsars in our galaxy similar to the Hulse–Taylor system made it possible to estimate, albeit very crudely, that the rate of final

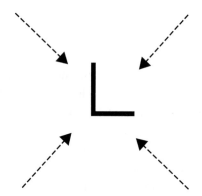

Fig 8.5 An interferometer is deaf to any gravitational waves approaching it parallel to the ground at an angle of 45 degrees relative to the arms.

inspirals of such systems in our galaxy could be of the order of one every 100,000 years. Obviously this is not a rate of occurrence on which to build a thriving career in gravitational wave detection. And indeed this fact was the Achilles heel for Joe Weber's attempts to detect gravitational waves. His detectors might have been capable of sensing waves from a nearby binary neutron star inspiral in our galaxy, but the rate of such events was far too small.

However, the strength of the gravitational wave signal received on Earth is inversely proportional to the distance of the source. If the waves from a source at a given distance have a certain strength, then waves from an equivalent source twice as far away are half as strong; waves from such a source ten times farther way are one tenth as strong, and so on. So if one could build an interferometer sensitive enough to detect waves from such a distance that a sphere surrounding the Earth of that radius contained about a million galaxies, then all of a sudden the rate of detectable neutron star inspirals could be a few (maybe up to ten) per year. A graduate student could even get a Ph.D. out of detections that frequent.

It is important to remember that a laser interferometer differs from a normal telescope in a crucial respect. A telescope can see light coming from only a very narrow range of directions and therefore has to be pointed toward a potential source. It is blind to all other directions. By contrast, a laser interferometer can hear gravitational waves coming from essentially *every* direction. There are only four source directions to which the interferometer is "deaf." If the wave approaches the interferometer traveling parallel to the ground at the site, but making an angle of 45 degrees with respect to the arms, the interference pattern when the laser beams recombine will not change at all (see Figure 8.5). But if the wave is approaching at just 7 degrees off from those special directions, the interferometer will hear the signal with 25 percent efficiency. At 15 degrees, the efficiency is up to 50 percent. Only a tiny number of signals unlucky enough to arrive from the vicinity of those four special directions would be missed. So if the designers and builders of LIGO and Virgo could reach the required sensitivity to hear inspirals from the nearest million galaxies (roughly a million times more sensitive than Weber's resonant bars), then one could imagine a decent rate of detections.

Another reason that the neutron star inspiral took hold as the most popular potential source is that the gravitational wave signal could be calculated very accurately from Einstein's theory. General relativity is a notoriously difficult and complicated mathematical theory, but fortunately the problem of two very small bodies orbiting each other was one that could be attacked by a number of methods. In one method, the equations of the theory could be replaced by an approximate set of equations which could be solved for the orbital motion and the gravitational waves emitted. This approximation could then be systematically improved in a step-by-step procedure. The results of these calculations were enormously long formulae that would typically span two or three pages of paper. But using these formulae and a simple laptop computer or tablet, one could generate a very accurate prediction for thousands of cycles of the inspiral part of the gravitational signal in a tiny fraction of a second. The last few cycles of a characteristic inspiral signal are displayed in Figure 8.4. In another method, the exact equations of the theory were formulated in a way that they could be solved numerically using large computers. This method was particularly important for accurately predicting the merger part of the signal, also shown in Figure 8.4. By the early 1990s, as inspirals were being recognized as the leading potential source for detection by the interferometers, Kip Thorne realized the importance of having accurate predictions of the waves in advance, and he urged theorists to begin the arduous task of developing these methods for predicting the wave shapes. Progress was slow at first, but by the time LIGO began its advanced run in September 2015, the tools for making accurate predictions were in place.

Why did this matter? The answer is a version of the child's puzzle that requires spotting a character named Waldo in the middle of a page of hundreds of cartoon characters; this version is called "Where's Nico?". If you were asked to spot a certain Nico in a teeming crowd of random people, knowing that Nico is a man would not be very helpful. But knowing that he is wearing a purple shirt with a black vest and a white *gaucho* hat would be very useful in finding him. The more you know about a signal, the easier it is to find it, even if it is buried in a lot of noise. In the same way, one could compare a library of predicted wave shapes (similar to the wave shown in Figure 8.4) to the output of the detector to see if there was a match. This would be done, not by eye, which is notoriously unreliable, but using sophisticated and extremely fast data analysis algorithms. Also, since the wave shape depends on things like the masses of the two bodies, that wave from the library that gives the best match to the data provides information about the system that produced the gravitational wave in the first place.

The neutron star inspiral idea was so popular that if you had asked anybody in the field just prior to the start of the Advanced LIGO run in 2015 what would be the most likely first detection, a large majority would have predicted a binary neutron star merger. The authors of this book certainly would have predicted that, though neither of us felt strongly enough to put serious money or seriously expensive wine on the table. But according to a Danish proverb (also attributed to Danish physicist Niels Bohr), prediction is difficult, especially about the future. Our predictions were wrong.

Not only was the first signal detected that of a binary *black hole* inspiral and merger, but the masses of the two black holes were completely unexpected. Until then, all the observational evidence and theoretical modeling on black holes pointed to two basic classes of black holes. The first was the class of *stellar mass* black holes, with masses between 6 and 15 solar masses, the classic example being the 10 solar mass black hole in Cygnus X-1. The second was the class of *massive* black holes, with masses between a hundred thousand and several billion solar masses, residing in the centers of galaxies, such as Sgr A* or the black hole in M87. We encountered these black holes in Chapter 6. A third class of *intermediate mass* black holes, between 100 and 100,000 solar masses, has been proposed, but the evidence is still not very solid.

Suffice it to say that black holes with masses of 36 and 30 solar masses were *not* expected. This was the set of values that gave the best fit between the theoretical wave shapes and the observed wave shapes during the inspiral phase. From that portion of the signal (the left-hand part of Figure 8.2), we can infer that the black holes were separated by roughly 700 kilometers, each revolving around the other at about one fifth of the speed of light, emitting waves at a frequency of about 50 hertz. By the time they collided and merged, each was moving at roughly a quarter of the speed of light. Using those masses and the equations of general relativity, and recalling that the overall strength of the waves decreases inversely with distance, one can then calculate how far the source would have to be so that the overall size of the wave agreed with the measured size shown in Figure 8.2. The answer turned out to be 1.4 billion light years. The final waves were emitted 1.4 billion years ago, around the time when the first green algae were forming in the oceans of the primitive Earth. Even though the two black holes were extremely far away, their larger than expected masses made the waves still "loud" enough at Earth for LIGO to detect them in 2015.

When two black holes merge they form a single black hole in a process that pictorially is similar to the way two soap bubbles merge into one big soap bubble. The main difference is that the "surface" of each black hole, known as the event horizon (see Chapter 6), is not made of any material such as soap, but is a surface where the warpage of spacetime has specific characteristics (such as allowing you to go in but never come back out). But just as the newly formed soap bubble may initially have a convoluted shape, the black hole remnant is a highly distorted beast, unlike the simple pictures you might see that represent black holes as dark spherical objects. The spacetime surrounding it also has bumps and distortions that vibrate and generate additional gravitational waves as the distorted black hole rotates. The frequency of the waves is related to the mass of the black hole, just as a bell whacked by a hammer emits sound of a specific tone. Eventually, friction within the metal of the bell reduces the vibrations and the bell goes quiet. But for the black hole "bell," these waves, called "ringdown" waves, carry the energy of the vibrations away so effectively (some waves go into the hole as well) that the black hole goes quiet after only a few cycles.

Unfortunately, the gravitational waves detected in 2015 were not loud enough during this ringdown phase to extract the ringdown frequency and decay time directly from the data, but something else could be done. The LIGO collaboration was able to extract the masses of the black holes from the earlier part of the signal, as we described earlier. Using this, together with numerical simulations of the Einstein equations, scientists were then able to predict that the mass of the final black hole was 63 solar masses, which is roughly consistent with the frequency of the ringdown part of the signal shown in Figure 8.2. Incidentally, those final cycles of radiation also provided evidence that the final object *is* a black hole. Other stellar objects, such as neutron stars, white dwarfs or ordinary stars, also vibrate at specific frequencies (the Sun has its own set of modes of oscillation), but their frequencies and decay rates are nowhere near the values inferred for the remnant of GW150914.

If you noticed a discrepancy between the initial total mass (66 solar masses) and the final mass (63 solar masses), you are correct. Three solar masses were converted into the energy of the outward-flowing gravitational waves, and most of this happened during the two tenths of a second of the signal detected by LIGO (see Figure 8.2). At its peak this represented a rate of energy output that is larger than that of all the stars in the observable universe combined! The energy emitted was the equivalent of a decillion (i.e. a million billion billion billion, or 10^{33}) 1 megaton hydrogen bombs. Interestingly, while stars and hydrogen bombs convert one form of matter into another (mainly hydrogen into helium), with the mass difference being converted into energy, here there is no matter. Whatever matter was involved in forming the black holes is inside their event horizons and can no longer affect the outside world. Instead, the conversion is from the mass imprinted on the spacetime curvature surrounding each hole into the energy of the spacetime ripples propagating outward.

In fact, using the characteristics of the detected wave and the distance to the source, one can estimate the total energy carried by the waves in all directions (some of which might conceivably have been detected by a LIGO in a galaxy on the far side of the universe). That energy turns out to agree reasonably well with Einstein's famous formula $E = mc^2$, where m is the 3 solar mass difference between the initial and final masses of the source, and c is the speed of light. That iconic formula, so well verified in the laboratory, also holds on cosmic scales!

You can think about this loss of mass–energy in another way. We saw in Chapter 5 that the orbital period of binary pulsars decreased as gravitational waves took energy away from the system. In that case, the rate of change of orbital period was minuscule, roughly 70 microseconds per year. But for the LIGO observations, the period change was so fast that one could see it in the 0.2 second data stream by eye (Figure 8.2)! Using the same formula that relates energy loss to period change for the binary pulsar, one could show that the the amount of energy lost during the inspiral in order to cause the observed period change was perfectly consistent with the 3 solar mass difference.

Everything about the event GW150914 supported key predictions of general relativity: gravitational waves exist, and carry energy away from the source. We knew this indirectly already from binary pulsars, but a direct detection and confirmation was essential. Binary black holes exist. There was already abundant observational evidence for black holes using electromagnetic radiation from material swirling or stars orbiting around them (see Chapter 6), but this was the first evidence using gravitational signals alone. Until then there was no observational evidence for binary black holes. This was the first, but it would not be the last. Newly formed, distorted black holes settle down by emitting "ringdown" gravitational radiation. This was confirmed.

GW150914 provided another important test of general relativity. The theory predicts that gravitational waves travel at exactly the speed of light, and, as with light traveling in vacuum, the speed is independent of the frequency of the waves. But during the inspiral, the frequency of the emitted waves changes dramatically, yet Einstein's theory predicts the waves always travel at the speed of light. In some modified theories of gravity, however, this is not the case.

One example is a class of theories developed in an effort to explain the fact that the universe appears to be expanding faster with time rather than slowing down with time, as standard general relativity predicts (see page 12). In these theories the "particle" associated with gravitational waves is given a small mass. Just as light can be thought of both as a manifestation of Maxwell's electromagnetic waves and as a quantum mechanical particle called the photon, vibrations in gravity can also be thought of as a manifestation of Einstein's gravitational waves or as a quantum mechanical particle called the graviton. But while the particle aspect of light has been abundantly demonstrated and even finds practical applications in things like solar panels, gravity is so weak that we will never be able to measure the particle aspect of gravity directly. If gravitons had a mass, then they would travel more slowly than light, with their velocity dependent on the wavelength of the waves associated with them. In particular, waves of long wavelength would travel slower than waves of short wavelength. Figure 8.6 illustrates what happens: waves emitted in the final part of the inspiral and merger travel a bit faster than waves emitted in the early inspiral. This then allows the merger waves to "catch up" with the inspiral waves in their long, long journey from the source to the Earth. The waves received by a detector would therefore be compressed or squashed a bit in time, compared to the wave that was emitted. But the signature of such a distortion was not seen at all in the LIGO observations, once again confirming Einstein's predictions in a spectacular way. The LIGO observation places a bound on how large the mass of the graviton could be, as otherwise, if it were any larger, they would have been able to detect its effects. The bound says that the mass of the graviton must be smaller than one part in ten octodecillion kilograms, in other words one divided by ten followed by fifty-seven zeroes! General relativity predicts that its mass is exactly zero, in agreement with the measured bound.

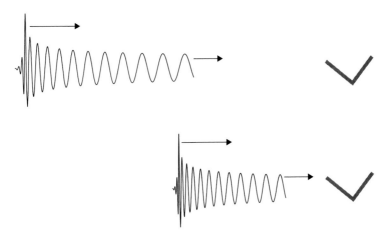

Fig 8.6 If short-wavelength gravitational waves travel more quickly than long-wavelength waves, then the later "merger" waves will catch up with the earlier inspiral waves, so that the wave received by the interferometer is slightly squashed in time compared to the emitted wave. No such effect was detected in the data, placing a very tight upper limit on the mass of a hypothetical "graviton."

Despite the string of experimental successes for general relativity that we have been describing in the first part of this book, experimental verifications like these matter. The reason is that Einstein's theory had not yet been seriously tested in extreme gravity situations, where gravity is very strong and rapidly changing. At the same time, observations of the universe as a whole have revealed apparent anomalies such as the accelerating expansion of the universe. The combination of these two facts has led theorists to propose that some of these anomalies could be resolved by modifying general relativity. And the potential outcome of these tests could be dramatic: any measurement that demonstrates and confirms a deviation from Einstein's predictions could direct us toward a resolution of observational anomalies, potentially leading to a theory of gravity that is "beyond Einstein."

The "discovery" event GW150914 was not the only one recorded during that first observing run, called O1 by the LIGO collaboration, which ran from mid September 2015 until 19 January 2016. The second event came on 12 October, but it was so weak that the statistical probability that it was a fluke was just once every three years, much larger than the probability of less than once every 200,000 years for the discovery event. Because of this, the LIGO collaboration decided not to announce this event as a proper detection, but rather as a "candidate" event which could not be confidently confirmed as an actual gravitational wave. A later reanalysis would promote this to a reasonably confident detection of a 23 and 14 solar mass black hole merger. The third signal was detected in the early hours of 26 December 2015, Greenwich time, although it was still Christmas day in the US. It was called the "Boxing Day event," after a tradition in England and Commonwealth countries of giving tradesmen, postal workers or staff a box containing cash or a small gift on the day after Christmas. This event was quite similar to the first one in that it was produced in the merger of black holes, located roughly the same distance from Earth, but with smaller masses (14 and 7 solar masses). Consequently, this event was not as loud as the first one, but still loud enough to be statistically significant.

The events we just described are the only ones LIGO detected during its first observing run. Following some improvements in the sensitivity of the detectors, the second observing session, O2, ran from November 2016 to the end of August 2017, and the haul of detections was impressive. A high-mass binary black hole inspiral (31 and 20 solar masses) was detected on 4 January 2017, and a moderate-mass inspiral (11 and 8 solar masses) was found on 8 June. Four more black hole inspirals were detected by LIGO during O2 between late July and the end of August, including an event with two huge black holes, weighing in at 50 and 34 solar masses, producing a final monster hole of 80 solar masses. This source was also the most distant by far, at nine billion light years. The other three detections were quite similar to the discovery event GW150914 in mass, distance and other characteristics.

Ironically, these four detections made during the summer of 2017 were not announced until December 2018, because of what occurred during the fateful week of 13 August 2017. Two detections made three days apart during that week were so exciting that everything else was pushed aside in order to focus on these finds.

As we discussed in Chapter 7, the Virgo gravitational wave detector in Italy was about a year behind LIGO in the various stages of construction, commissioning and upgrading. But on 1 August it joined LIGO in a three-way session, scheduled to last for three weeks, until the end of O2. And on Monday 14 August 2017, the three instruments heard the gravitational waves emitted by another binary black hole merger, with masses in the 25 to 30 solar mass range. But the triple detection allowed for something new: pinpointing the location of the source in the sky.

How can these detectors determine where the waves came from? Just as in navigation using GPS, discussed in Chapter 2, the answer is triangulation, exploiting the arrival time of a signal. In GPS, it is the arrival time at the user of signals from multiple GPS satellites that allows the user to determine her location. In gravitational wave triangulation, it is the arrival of the signal from a single source at different detectors. For two LIGO detectors, we saw in Figure 8.3 how measuring the difference in arrival times at the two sites allows you to determine that the source lies somewhere on a circle on the sky (except for the special case where the source lies precisely along the line joining the two detectors). But if there is a third detector, such as Virgo, then in exactly the same manner, the difference in arrival time between Virgo and, say LIGO-Livingston, gives a second circle on the sky (the time difference between Virgo and LIGO-Hanford gives redundant information).

There are three possibilities for these two circles. The first is that they do not intersect at all, in which case the event would have to be rejected as a candidate gravitational wave. This is analogous to the case where the time difference between the arrival at two sites is *larger* than the light travel time between the sites. The second possibility is that they intersect in two places, as illustrated in Figure 8.7. In this case, since the source must reside somewhere on both circles, it must therefore be at one of the two intersection points, either *A* or *B*. The third is a very special and lucky case in which the circles just touch each other at a single point, giving a single location for the source on the sky. In reality, these circles are really bands of some width, reflecting the uncertainties that are inevitable in noisy and imperfect data. And exploiting additional details of the response of the detectors to an impinging wave, it is possible to exclude parts of each circle as being less likely to correspond to the source's location. Thus, while data on the prior LIGO-only detections could

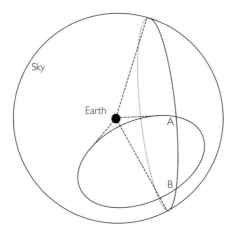

Fig 8.7 The arrival-time delay between two detectors fixes the source to lie on a circle in the sky. The delay between one of those and a third detector fixes the source on another circle. If the circles intersect, the source must lie at one or other intersection point (A or B).

confine the sources at best to large elongated banana-shaped regions on the sky, the data on the LIGO–Virgo source GW170814 confined it to an oval region in the southern sky about the size of a major league baseball held at arm's length.

Three days later, on 17 August, another signal passed the Earth, reaching Virgo first, then LIGO-Livingston 22 milliseconds later, and LIGO-Hanford 3 milliseconds after that. This signal was quite different from all the black hole inspiral signals detected to that point. Instead of a very short, rapidly changing chirp–merger–ringdown signal, as shown in Figure 8.2, the signal was detected for a whopping 100 seconds, and appeared pretty boring, something like the signal shown schematically in Figure 8.8. In fact, Figure 8.8 illustrates only about the final quarter of a second of the signal (to picture the whole signal you have to imagine it continuing about 200 page widths to the left). The signal was a regular undulating wave, suggesting that the source is a binary system. Its frequency was roughly the same as for the black hole mergers, around 100 hertz, suggesting that the bodies are revolving around each other very fast. But compared to the black hole signals, where the changes in

Fig 8.8 Gravitational wave of a neutron star binary inspiral. Only the last quarter of a second of the wave is displayed. The observed wave GW170817 lasted 100 seconds.

size and frequency of the wave could be seen over a few orbits, here the orbit-by-orbit changes are minuscule, suggesting that the rate of leakage of energy into gravitational waves is tiny. This indicates that the masses of the two bodies are much smaller than the masses in the black hole inspirals. All of this pointed to an inspiral of two neutron stars. The event was denoted GW170817.

In addition, while the signal was detected in both LIGO instruments, it was *barely* detected by Virgo. While it was possible that the Virgo detector was somehow malfunctioning at that particular moment (which would have been strange, since it worked perfectly three days earlier), the more likely explanation was that the source was at a location in the sky close to one of the four directions for which Virgo is "deaf" (see Figure 8.5). Because of that unlucky alignment, the Virgo interferometer did not respond as strongly to the signal as it might otherwise have. This useful information, combined with the circle in the sky inferred from the arrival times at the two LIGO detectors, gave a decent sky location for this source.

Meanwhile, orbiting 534 kilometers above the Earth, detectors on board the Fermi Gamma-ray Space Telescope detected a burst of gamma rays in its routine sweep of the sky. The gamma rays arrived 1.74 seconds after the end of the gravitational wave signal. Fourteen seconds later, even before the automated software at LIGO and Virgo had fully registered what they had detected, Fermi issued an automated alert to astronomers worldwide (including the LIGO–Virgo team) so that follow-up observations could begin. This detection was denoted GRB170817 (for gamma ray burst). Forty minutes later, LIGO–Virgo issued its own worldwide alert, noting the near coincidence in time between the gravitational wave signal and the gamma ray burst. Five hours after the gravitational wave event, they had localized the source to be in a region of the sky about 30 square degrees in angular size, and at a distance of 130 million light years, good to about 20 percent. These observations meant that the source was somewhere within a three-dimensional cube in space containing about 49 galaxies. The Fermi observation was consistent with this, but was not accurate enough to pinpoint the specific galaxy from which the gamma rays had originated. About 12 hours after the initial detections, a team using the Swope Telescope at Las Campanas Observatory in Chile detected a counterpart signal in the visible band, and identified the host galaxy as NGC 4993.

The galaxy NGC 4993 is not particularly interesting. Discovered in 1789 by astronomer Wilhelm Herschel, it is an elliptical galaxy in the constellation of Hydra. Unlike the Milky Way or the Andromeda galaxies, which are rotating galaxies with spiral arms and a dense core, an elliptical galaxy is more egg shaped, with stars that orbit almost randomly around its center. NGC 4993 is about the size of the Milky Way, and like most galaxies it hosts a supermassive black hole at its center, which in this case has a mass of roughly 80 to 100 million solar masses. It also shows evidence that it merged with another galaxy about 400 million years ago. So, apart from hosting humanity's first ever detected neutron star inspiral and merger, NGC 4993 is a fairly run-of-the-mill galaxy.

Ironically, while all this was happening that August morning, Nico was hosting a workshop at the eXtreme Gravity Institute at Montana State University in Bozeman, called "eXtreme Gravity meets eXtreme Matter." The main purpose of this workshop was to bring experts together to discuss the science one would be able to extract once LIGO–Virgo detected gravitational waves from the merger of two neutron stars. This made it extremely difficult for half the attendees (who belonged to the LIGO–Virgo collaboration and had seen both alerts) to participate in the workshop discussions, as they had to abide by LIGO secrecy rules. How they managed to hold their excitement at bay and not spill the beans is a mystery. Those constraints would be lifted only months later, after the event was confirmed as a true detection.

What followed the LIGO–Virgo and Fermi alerts was one of the most extraordinary observational campaigns in the history of astronomy.

Over the next thirty days, follow-up observations were made in every band of the electromagnetic spectrum, using telescopes on the ground and in space. Thirteen separate teams made gamma ray and X-ray observations. The Hubble Space Telescope made observations in the ultraviolet, visible and infrared. Thirty-eight teams observed in the visible band, while twelve worked in the infrared band. Fifteen teams made observations in the radio band. Three teams even looked for signals of neutrinos (not surprisingly, given the distance of the source, they saw none). On 20 October 2017, *Astrophysical Journal Letters* published a summary of what all the observations had yielded during those first two months in what was being dubbed "multi-messenger" astronomy. The paper had 3,500 authors, of whom 1,100 were in the LIGO–Virgo collaboration and the rest were associated with astronomical projects. Virtually every astronomy department or center in the world was represented in the 953 listed institutes. A year after the detection of GW170817 and GRB170817, observations of the electromagnetic radiation from the source continued in many wavelength bands, and almost a hundred papers had been published in a range of astronomy journals, along with around fifty in physics journals.

We might be forgiven for pointing out a certain irony in all this. In the early 1990s, when the US government was trying to decide whether to give the go-ahead for major funding to begin the construction of LIGO, many prominent American astronomers lobbied vigorously against it. The astronomers' arguments roughly boiled down to some combination of: it won't work; it will only detect gravitational waves, which we already know exist (see Chapter 5); it's just a physics experiment with nothing to do with astronomy; and it is too expensive. One astronomer, J. Anthony (Tony) Tyson, reported that an informal poll he conducted in early 1991 of seventy astronomers ran four to one against LIGO. In March 1991, Cliff testified in favor of LIGO before a US House of Representatives Science subcommittee, alongside Tyson who, while supportive of LIGO in general, testified against construction funding at that time. What a difference a detection makes!

What did we learn from this multi-messenger source? Detailed analysis of the gravitational wave signal indicated that the masses, around 1.5

and 1.3 solar masses, were quite consistent with those of known neutron stars. The gamma ray burst showed that it could not have been two black holes, because you need hot matter to get such high-energy radiation, and black holes are pure spacetime. A "mixed merger" of a neutron star and a black hole could not be definitively excluded, although it was hard to imagine where such a low-mass black hole would come from.

The association of a neutron star merger with a burst of gamma rays resolved a long-standing mystery. Gamma ray bursts have been observed and studied since the 1960s, when the US Vela satellites accidentally detected the first bursts. These satellites had been deployed in the middle of the Cold War by the US military to investigate whether the Soviet Union was testing nuclear weapons in space. Gamma rays are a byproduct of such nuclear explosions, and indeed the Vela satellites detected many bursts of gamma rays. But the bursts did not have the characteristics of those emitted by nuclear bombs, and appeared to be coming from far outside the solar system.

Within a few years, the study of these mysterious gamma ray bursts exploded. In 1991, the Compton Gamma Ray Observatory (CGRO) was launched, and over the next nine years it observed and localized around 2,700 gamma ray bursts (almost one per day), finding that they were not coming from any preferred direction, and thus suggesting an extra-galactic origin. Moreover, astronomers deduced that the bursts came in two rough classes based on their duration: "short" bursts had an average duration of about 0.3 seconds, while "long" bursts lasted 30 seconds on average. Although the short bursts comprise only about 30 percent of all observed gamma ray bursts, they were particularly interesting. Astronomers realized that there was no correlation between these short bursts and supernovae, ruling out the latter as possible progenitors. They also realized that most of the short bursts came from elliptical galaxies where there is an underabundance of massive stars, which are needed for supernovae.

Theoretical arguments made the short bursts even more intriguing. Imagine that the source that produces these short bursts of gamma rays is a ball or blob of matter of some size. Imagine that the burst is caused by an explosion of the blob, producing a flash of light that is directed toward us.

The duration of the burst must be related to the finite extent of the source: we first see the light emitted from the region of the source closest to us, and later we see radiation emitted from other regions farther away from us. Since the entire duration of the burst is about 0.3 seconds, and since gamma rays travel at the speed of light, we conclude that the emitting region must be of the order of the time duration multiplied by the speed of light, which comes out to roughly 100,000 kilometers, or eight times Earth's diameter. And because of the tremendous energy produced in these bursts, there had to be an enormous amount of matter contained in such a small region of space. Whatever produced these short bursts had to involve dense compact objects such as neutron stars, as Russian physicist Sergei Blinnikov and collaborators had theoretically predicted back in 1984.

By 2005, astrophysicists began to suggest that the short bursts were the result of either neutron stars merging with each other or of a neutron star merging with a small black hole. But there was no way to prove this, since no light would be detected from the merging pair before the short gamma ray burst started. The LIGO–Virgo observations provided the missing piece of the puzzle, unequivocally proving that at least one of the possible progenitors of short gamma ray bursts is the merger of neutron stars. This observation also validated other ingredients of the model, such as the fact that the bursts are emitted in a very narrow cone, which we detect on Earth only if the cone happens to be pointing in the right direction. This, in turn, suggests that many more short gamma ray bursts must be occurring in the universe, but with their emissions beamed in cones that do not point toward Earth. Gravitational wave observations will be able to determine precisely how many such events occur in the universe, since gravitational waves are not emitted narrowly in a cone, but rather more or less equally in all directions.

These short gamma ray burst models also claimed to answer a question asked by anybody who has ever bought a wedding ring or examined the insides of a cell phone: where does gold come from? Today we have a very good idea of the origin of the key elements in nature. About three minutes after the big bang, around 20 percent of the primordial hydrogen was converted via nuclear fusion to helium, along with a sprinkling of lithium, a process known as big bang nucleosynthesis. Stars continue the process, converting more of their hydrogen into helium, but also extending the fusion process to elements such as carbon, nitrogen and oxygen, so crucial for the life forms that inhabit Earth. Very massive stars also produce those elements, but when they explode as supernovae, they produce elements up to the so-called "iron group" (iron, manganese, cobalt, nickel) and spew them into interstellar space, later to be incorporated into planets such as Earth. Unfortunately it is not so easy to produce elements heavier than iron. Many elements above iron in the periodic table are unstable, decaying to other elements by various radioactive processes, and they all have many more neutrons than protons in their nuclei. Nuclear physicists had developed a set of chain reactions, known as "r-process nucleosynthesis" ("r" meaning rapid, not the most imaginative of names), that could in principle produce heavier elements in the right proportions, but they all required that the processes take place in an extremely neutron-rich environment. Normal stars and supernovae utterly failed to produce the right environment. But neutron stars are almost completely made of neutrons (with a small contamination of protons and electrons), and so proponents of the neutron star merger model for short gamma ray bursts argued that this would be the right environment for the r-process.

One class of models for what happens after two neutron stars merge came to be called "kilonova models." This is a variant of the word "supernova," but it has nothing to do with the explosion of a massive star. According to these models, when neutron stars merge, they will spew a few hundredths of a solar mass of material into space at about a few tenths the speed of light. Some of this material will fall back onto the remnant, maybe forming a disk of material that is slowly swallowed or "accreted" by the remnant. But some of the ejected material will have enough of an initial velocity after the collision that it will escape altogether, creating a cloud of very hot and very neutron-rich material. As the cloud expands and cools, the r-process produces elements like gold, platinum and silver, as well as heavier elements in the lanthanide part of the periodic table (elements that are crucial for computers, cell phones and batteries).

And this is precisely what optical and infrared telescopes observed when following up the gravitational wave detection. In fact, the electromagnetic radiation detected from the radioactive decay of material in the hot cloud expanding from the merger site suggests that about fifty times the mass of the Earth was produced in silver, a hundred times the mass of the Earth in gold and five hundred times the mass of the Earth in platinum, a mere second after the merger.

Astronomer and science TV host Carl Sagan once said that we are made of star dust. He was referring to the elements forged through nuclear fusion in stars and supernovae. But now we know that we are not just that. A little part of us also contains neutron star dust (not too much, as that would be toxic). What's more, neutron star dust (in the form of gold, platinum and silver) is forged into jewelry that routinely adorns our bodies. We even sometimes bind ourselves to each other through rings of neutron star dust.

We close this chapter with an extraordinary test of Einstein's theory provided by GW170817 and GRB170817. We already learned from the black hole inspirals that the speed of gravitational waves is independent of wavelength, in accord with general relativity, but we learned nothing about the actual value of that speed. That is because only gravitational waves were received from those inspiral events, and since we do not know exactly when the signals were emitted, there is no way to calculate the speed of the waves. General relativity predicts that the speed of gravitational waves is *exactly* the same as the speed of light. The nearly simultaneous arrival of the gravitational waves and the gamma ray burst proved that they are the same to incredible precision. The argument goes like this.

The LIGO–Virgo collaboration compared the time at which the peak of the gravitational waves arrived at the detectors to the time at which gamma rays arrived at the Fermi satellite. After correcting for the altitude and location of the satellite and for the radius of the Earth, the scientists concluded that the gamma rays arrived about 1.7 seconds after the peak of the gravitational wave train. This delay is presumably caused by the fact that the gamma ray emission did not start when the neutron stars first touched each other, which does coincide with the peak of the gravitational wave signal, but instead the gamma ray emission originated in whatever violent explosion followed the merger. The details of that explosion are still an area of active research, but different models suggest a delay between 1 and 10 seconds.

Let us assume that the gamma rays were emitted at *exactly* the same time as the peak of the gravitational waves, that is, roughly at the time the neutron stars first touched. If so, then we would attribute the delay in the gamma ray arrival entirely to gravitational waves moving more quickly than the gamma rays. How much more quickly would they have traveled? The travel time of the gravitational waves is simply the distance traveled divided by their velocity, and similarly the travel time of the gamma rays is the same distance divided by the speed of light. Remember that the distance over which this little race occurred is 130 million light years. The difference in these travel times must equal the 1.7 second time delay measured, which then allows us to estimate that the gravitational waves were faster than the gamma rays by at most about three quarters of a millimeter per hour. Alternatively, let us assume that the gamma rays were emitted 10 seconds *after* the gravitational waves. In this case the faster gamma rays would have narrowed the gravitational waves' head start to 1.7 seconds, requiring them to be faster by about 3 millimeters per hour. Comparing these two estimates to the speed of light, a million billion millimeters per hour, one sees that the permitted speed difference is truly minuscule: smaller than parts in 10^{15}! As we discussed earlier, many of the latest attempts to account for the accelerated expansion of the universe and the effects attributed to dark energy by invoking an alternative gravitational theory *require* that the speed of gravitational waves be different from that of light. The single observation from GW170817 and GRB170817 that this is not the case forced theorists to throw a large heap of such theories immediately into the trash. Conversely, if we assume that general relativity is correct, and thus that the two speeds are identical, then the 1.7 second delay between the arrival of the two signals corresponds precisely to the delay between the emission of the two signals, a fact that is already proving very important in sorting through the many complex models for how the gamma rays were generated.

In the end, most of us were wrong in our predictions that neutron star mergers would be the first events detected by LIGO–Virgo. But hey, ninth place isn't so bad! Those unexpectedly massive black hole inspirals were loud, and the rate of such events is obviously higher than people expected. The third observing run of LIGO–Virgo began on 1 April 2019 at even better sensitivity, with the expectation of more detections of black hole mergers, neutron star mergers, and maybe even mixed mergers of a black hole and a neutron star. As we end this chapter we don't want you to get the impression that this is it. The ground-based interferometers will also be searching for other sources of gravitational waves, and completely different kinds of detectors are in operation or under development, including one destined to go into space. The future is loud for gravitational wave detection, and we can look forward to hearing many more movements in the symphony of the universe.

CHAPTER 9

A Loud Future for Gravitational Wave Science

After reading Chapter 8, you may think that there is nothing else to learn from gravitational waves. But the detections of 2015–2017 are only the beginning. After a two-year hiatus to make additional improvements at the LIGO and Virgo observatories, a third observing run, called O3, began on 1 April 2019. Right away, new detections were made on 8 April, 12 April and 21 April. By the middle of August 2019 there were around eighteen binary black hole mergers, two binary neutron star mergers, and, on 14 August, a possible black hole–neutron star merger. There were even a few events that were later retracted when additional analysis revealed them to be either instrumental or terrestrial in origin. In contrast to the extreme secrecy that surrounded the early detections, the LIGO–Virgo collaboration has gone fully public, now announcing detections as they come in. In fact, you can download a "gravitational wave" app to your smartphone that will send you an alert every time a sufficiently credible signal is detected. You can select a generic alert tone, or you can choose a tone that sounds like a gravitational wave "chirp." The reason for this openness is to allow astronomers to react more quickly, in the hope of finding electromagnetic counterparts to the gravitational wave events. By the time you read this paragraph, it will almost certainly be hopelessly out of date, with a regular stream of recorded signals from binary black holes, binary neutron stars, mixed mergers, or possibly something entirely new and unexpected.

But that is not all. There is an ambitious effort to take gravitational wave physics to the next level, and it begins in Japan. Mount Ikeno is near the city of Hida, in the center of Japan. About 1,000 meters below the mountain's surface is the Mozumi mine, owned by the Kamioka Mining and Smelting Company, which operated for over a century. Today, the mine is no longer operational, at least not for the extraction of minerals. Instead, it houses some of the most sophisticated physics experiments in the world. Super-Kamiokande, for example, is a detector built to observe and characterize high-energy neutrinos, those tiny subatomic particles that travel very close to the speed of light. They can be created in the Sun as a byproduct of nuclear fusion, or in the atmosphere when other high-energy particles, emitted in supernovae and other events, collide with atoms such as nitrogen and oxygen. Super-Kamiokande also looked for possible decays of the proton, one of the elementary building blocks of atoms. It found none, and provided a lower limit on the proton's lifetime of around 10 billion trillion trillion years (that's a 1 followed by 34 zeros).

The Mozumi mine is also home to a new gravitational wave detector that is beginning operations: the Kamioka Gravitational Wave Detector, or KAGRA. This detector is similar to LIGO and Virgo, with two perpendicular arms that are 3 kilometers long. Unlike LIGO and Virgo, however, the KAGRA detector is deep underground. And unlike the LIGO/Virgo mirrors, which are at room temperature, KAGRA's mirrors will be cooled to around 20 kelvin (-253 degrees Celsius or -423 degrees Fahrenheit). But why?

Going underground helps to diminish two important sources of noise: seismic noise and "gravity noise." The Earth is continuously shaking, even if we can't always feel it move under our feet. These seismic vibrations originate in erupting volcanoes, the grinding of tectonic plates against each other, or even man-made explosions. They travel in all directions on the surface of the planet and through the interior. Many of the interior waves are quenched because of the high densities and the molten core inside the Earth. The strongest vibrations travel along the

Earth's surface. Isolating the mirrors of LIGO and Virgo from these external vibrations was a major undertaking, but despite those sophisticated efforts some of these vibrations still get through, forcing the mirrors to shake a tiny bit. These vibrations are especially troublesome at the lower frequencies, say 10 hertz and below, getting in the way of our ability to detect the low-frequency vibrations due to gravitational waves. So, placing a gravitational wave detector deep inside a mountain is a way to protect the instrument from surface seismic vibrations.

Similarly, gravity above and around the instrument is not constant. Imagine a clump of cold, dense air passing over an interferometer on the surface (left panel of Figure 9.1). Because the mirrors are several kilometers apart, the mass in the clump can gravitationally attract the mirrors differently, inducing small motions that act as a background noise. You might say that such effects must be ridiculously small, and they are. But they can still be larger than the ridiculously small motions induced by the gravitational wave we want to detect. There is no way to shield gravity, but if you put the interferometer deep underground (right panel of Figure 9.1), you add distance between it and the clump. Since the force of gravity falls off inversely with the square of the distance to the source of the perturbation, this helps to mitigate the effect. Because everything with mass produces gravity, including people driving cars and even the compressions associated with seismic waves, gravity noise is a problem, and isolating the detectors underground can help reduce these effects.

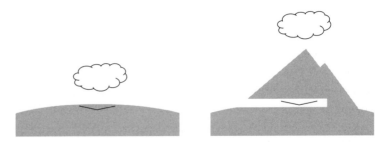

Fig 9.1 Left: The gravitational attraction of a dense cloud can move the mirrors in an interferometer on the ground. Right: For an interferometer such as KAGRA deep under a mountain, the cloud is farther away, reducing its gravitational effects.

So why were LIGO and Virgo not built underground? The reason is cost. As every Bostonian who has lived through the "Big Dig" project knows, digging tunnels is an extraordinarily expensive and complex undertaking even for something as worthy as managing traffic. Every major science project has to balance the scientific return and the risk of failure against the cost and timetable for construction. Compromises in scientific capability are often necessary to reach a budget that has a chance of being accepted by the governments who will be spending the taxpayers' money. So going underground was never in the cards for LIGO or Virgo. But KAGRA had the advantage of the pre-existing tunnels and infrastructure (power, air handling, access) of the Kamioka complex, and while some excavation was necessary, it wasn't a deal-breaker.

The advantage of cooling the mirrors to low temperatures is related to an effect called Brownian motion. Robert Brown was a Scottish botanist, who in 1827 observed that tiny particles ejected by grains of pollen and suspended in water seemed to be in constant motion when observed under a microscope. While this phenomenon was initially thought to be caused by some mysterious "life-force," it remained largely unexplained until Einstein developed a statistical theory of heat in one of his celebrated 1905 papers. Einstein realized that the jittery motion had to be a consequence of collisions between the grains and the fast and randomly moving water molecules. Heat, in fact, is nothing but a manifestation of the kinetic energy of the motion of molecules. As a result, the surfaces of all the mirrors in an interferometer are in a constant state of undulation, and this introduces an uncertainty in defining the precise distance between the mirrors' surfaces as determined by the bouncing laser beams. In addition, no mirror reflects perfectly, and some of the laser light is absorbed by the mirrors, heating them up and adding to the thermal undulations. Both effects can be reduced significantly by cooling the mirrors to temperatures near absolute zero.

KAGRA's sensitivity to gravitational waves is expected to be comparable to LIGO's and Virgo's, perhaps even exceeding them at low frequencies where seismic and gravity noise dominate those detectors. But the Japanese project suffered from a number of delays. For example, in 2014–2015 some of the tunnels of the Mozumi mine partially flooded because of an excess of snow melting in the spring. This water was a problem not just because experimentalists would get their feet wet, but because it could contribute to gravity noise. The development of a plan to deal with this excess water delayed the construction of the detector. But in 2018 KAGRA's construction was finally completed and a first run (with mirrors at room temperature) was successful. In early 2019, the KAGRA team reported a successful test with the mirrors at 16 kelvin, and readied themselves for joint observing runs with LIGO and Virgo toward the end of O3, in 2019.

Meanwhile, a fifth gravitational wave instrument is under development, called LIGO-India. It will be built near the town of Aundh, a suburb of Pune in the Hingoli district of Maharashtra, southeast of Mumbai. The town is dedicated to the Hindu Goddess Shiva, the destroyer and transformer, and it hosts one of her twelve most important representations, or *Jyotirlingas*, in India. LIGO-India will consist of a replica of one of the LIGO interferometers, but located in India instead of in the United States. In fact, this replica was initially part of a smaller interferometer at the Hanford site of LIGO.

When we have been describing LIGO and Virgo we have been a bit sloppy in using the term "interferometer." The actual interferometer consists of the lasers, the beam splitter, the mirrors, the seismic isolation equipment, along with all the detailed instrumentation needed to make it all work. There is an equally important ingredient of the observatories, namely the enormous kilometers-long vacuum tubes created to house the interferometers. We tend to lump them together in the single word.

When LIGO was first being designed, it was decided to fabricate and install *two* interferometers at the Hanford site. The second interferometer was identical to the first in every way—identical lasers, beam splitter and mirrors—except that the end mirrors were installed at the two-kilometer mark, instead of at the end of each vacuum tube, at four kilometers. This system was denoted H-2, while the longer system was denoted H-1 (the Livingston system was called L-1). In fact, the evacuated tubes were sized to be able to accommodate multiple interferometers and multiple laser beams.

The reasoning behind this was redundancy and confidence. A gravitational wave arriving at the detector would induce a response in H-2 that would be exactly one half the size of the response of H-1, but would be otherwise identical. The seismic noise background in both instruments would be almost identical. In contrast, the Livingston detector, while containing identical instrumentation, lived in a different seismic environment, with its own vacuum system for the beam tubes, and the concern was making a confident detection of a wave in the face of the disparate sources of noise. The initial LIGO runs from 2002 to 2010 involved all three interferometers at the two LIGO sites. When the improved instrumentation for the advanced LIGO detectors was being built during the period 2008 to 2010, an advanced interferometer system was fabricated for H-1, H-2 and L-1.

But it turned out that the runs of the initial LIGO system were extremely successful. True, they did not detect gravitational waves, but they didn't really expect to, because initial LIGO was not quite sensitive enough. But they learned so much about the various noise sources and how they affected the detectors that the LIGO team began to ask, do we really need H-2? If not, what could one do with the instrumentation that was being built? In 2009, Jay Marx, an experimental high-energy physicist who had taken over the directorship of LIGO from Barry Barish three years earlier, raised the idea of offering it, more or less free of charge, to somebody else willing to construct the vacuum tubes and to provide all the needed infrastructure for a working gravitational wave detector. Although the National Science Foundation was initially skeptical, it eventually signed off on the proposal.

What could possibly possess the United States to give away roughly 80 million dollars worth of valuable technology? The answer was science. We have already discussed the importance of multiple detectors for identifying gravitational wave signals using Weber's principle of coincident detections. That was the reason for two LIGO observatories in the first place, and for building the Virgo detector in Europe and KAGRA in Japan.

But more importantly, we have seen how multiple detectors can make use of the various arrival times of signals to triangulate the sky location of the source (page 231). The more detectors that see the event and the farther they are from each other, the more precisely they can localize the source. However, all the current detectors, LIGO, Virgo and KAGRA, are firmly in the northern hemisphere and lie within 15 degrees latitude of each other, with the southernmost, LIGO-Livingston, at 30 degrees latitude, and the northernmost, LIGO-Hanford, at 46 degrees latitude (Virgo and KAGRA are at 43 and 36 degrees respectively). So, to a very crude approximation, all four detectors lie on the same imaginary plane cutting through the Earth.

This means that a gravitational wave approaching the Earth from an area near due north or due south would produce rather small time delays among the four detectors, and thus the errors in locating the source on the sky will be larger, despite having four detectors. But having a detector in the southern hemisphere, far from this "plane," would add critical additional time delay information. Analyses showed that such a detector would enhance the ability to localize the source in the sky by a factor of five to ten.

A factor of ten is a big deal. In Chapter 8, we explained how the coincident detection of a gravitational wave produced in the merger of two neutron stars in 2017 by three instruments (the two LIGO detectors and Virgo) allowed for the localization of the source in the sky to a moderate accuracy. This, in turn, allowed astronomers all over the world to point their telescopes to the right spot in the sky and find the light emitted following the collision. But we were lucky with that one. The neutron star merger occurred quite close to Earth (a mere 140 million light years away), so the number of galaxies within the region singled out by LIGO and Virgo was not too large. This allowed the first observers on the case to scan the galaxies one by one to identify the galaxy with unusual electromagnetic emission, and then alert others as to the precise location.

If the source had been farther away, then it would have been much more difficult to home in on the right galaxy containing the merger event. A spot of a given size on the sky that one wishes to search for represents a larger physical area the farther away one wishes to look (see Figure 9.2), and if the area is larger, then more galaxies will exist within the chosen

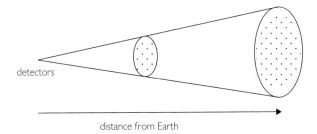

detectors

distance from Earth

Fig 9.2 The importance of good localization of gravitational wave sources on the sky. For a nearby source the region on the sky may contain only a handful of galaxies, making it possible for astronomers to scan them one by one in search of an electromagnetic counterpart. For a more distant source, the number of galaxies in the region increases, making scanning by telescopes more challenging. The narrower the cone, the smaller the number of galaxies to scan.

region. Thus, localizing the gravitational wave source more narrowly can be a big advantage, allowing telescopes to find the electromagnetic emissions much faster, catching the emitted light earlier in this rapidly varying process.

As a result, LIGO and the NSF specifically sought out a partner in the southern hemisphere to house the H-2 interferometer. The plan would be to put the interferometer equipment into storage, and then when the partner had constructed the large vacuum tubes and all the related infrastructure, the US would hand it all over to be installed. They first tried Australia, which had a very active gravitational wave effort, with many scientists already members of the LIGO–Virgo collaboration. There was even the 80 meter prototype interferometer AIGO near the town of Gingin, about 100 kilometers north of Perth, developed by researchers at the University of Western Australia. It had been used to develop advanced instrumentation for LIGO. But despite approval of the plan by the NSF and intense lobbying by Australian scientists and gravitational physicists worldwide, in October 2011 the Australian government declined the opportunity, saying that it did not have the money in its science budget for such a new endeavor.

A year later the US government approved an approach to India. Unlike the Australians, the Indian gravitational wave scientists did not have a home-grown interferometer prototype, but many of them were members of the LIGO–Virgo collaboration, and had spent time at the various sites. They quickly formed a consortium of enthusiastic researchers and began lobbying the relevant agencies of the Indian government to promote LIGO-India. The approvals slowly worked their way up the various government hierarchies. A temporary glitch in the process occurred when somebody in the Indian government realized that the detector would be measuring seismic signals continuously and that this data would be shared by the entire collaboration of researchers. Such data was viewed by the government as a matter of national security, because it was used to monitor whether the country's neighbor Pakistan was conducting nuclear testing, and therefore it could not be viewed by anybody outside India. Reaching a workable compromise on this issue set negotiations back by about a year. Finally, on 17 February 2016, Prime Minister Narendra Modi announced the government's approval of the project. Interestingly, and perhaps not entirely coincidentally, this was a week after LIGO's announcement of the first gravitational wave detection. Construction is expected to begin in 2020, with the first science collection expected roughly five years later.

Researchers in Japan and India are not alone in thinking of the future of gravitational wave physics. The LIGO–Virgo collaboration has been thinking hard about how to make the detectors in Hanford, Livingston and Cascina even more sensitive. The first suite of upgrades is planned for the beginning of the 2020s. One upgrade envisioned is to employ improved reflective coatings on the mirrors to reduce the amount of laser light absorbed by the mirrors themselves. This will help reduce the thermal undulations of their surfaces.

Another upgrade is to introduce a technique that laser scientists have been perfecting in recent years called "squeezing" of light. A laser beam consists of light particles or photons. When a photon bounces off a mirror, the mirror recoils a small amount, just as a billiard ball recoils when struck by another ball. This is an example of the conservation of momentum. But because of the disparity between the momentum of a tiny photon and that of the mirror, the recoil of the mirror is ridiculously small. (By now you may be used to the fact that "ridiculously small" can still be a problem for gravitational wave detection.) And since photons are fundamentally governed by quantum mechanics, Heisenberg's uncertainty principle introduces randomness in their behavior that leads to noise in the motion of the mirrors, especially at high frequencies. While Heisenberg's principle asserts, for example, that you cannot know both the position and velocity of a particle to arbitrary precision, it fully permits you to reduce the uncertainty in one variable, as long as you can live with a larger uncertainty in the other. With squeezed light, you can decrease the uncertainty in the aspects of the photon beam that make the mirrors recoil, while letting it increase in aspects that don't affect the mirrors. The result is to reduce this "photon noise" in the high-frequency regime. This was first demonstrated using the GEO-600 detector in Hanover, Germany in 2011, and is being implemented in the larger interferometers.

The next round of upgrades is planned for around 2025–2030, and will involve importing the KAGRA technology to cool the test masses down to a temperature of about 120 kelvin. Other improvements include new mirrors made of silicon and new suspensions for the mirrors that will better isolate them from seismic vibrations. When all is said and done, these new detectors should be twice as sensitive as advanced LIGO. A factor of two improvement in sensitivity doubles the distance to which sources can be detected, increases the accessible volume by a factor of eight (since volume increases as radius cubed), and thus increases the number of galaxies accessible to detectors by a factor of roughly eight!

The upgrades of the current LIGO and Virgo detectors, together with the new detectors in Japan and India, are being called second-generation detectors. They involve basically improving on the original "L"-shaped concept envisioned in the 1970s by Gerstenshtein and Pustovoit, Weber and Forward, and Rai Weiss. There is an ambitious plan being hatched for a "third"-generation detector of a quite different design, called the Einstein Telescope, or ET.

The Einstein Telescope is currently planned as an underground triangular interferometer, in which the angle between the arms is 60 degrees instead of 90. It turns out that there is no law that says that an interferometer *must* have arms at right angles to each other. The current interferometers have right-angled arms largely for historical reasons. The original table-top interferometers built by Michelson in the late nineteenth century to measure the speed of light and to try (unsuccessfully) to detect the motion of the Earth through a hypothetical "aether" had right-angled arms. But history is not the only driver for 90 degree arms; it turns out that at 90 degrees, the response of an interferometer to a given incoming gravitational wave is the largest it can be. However, if you make the angle smaller or larger than 90 degrees the response does not decrease by all that much. At 60 degrees between the arms, the response is still about 86 percent of the maximum possible (the actual factor is the sine of the angle between the arms).

But then, a very nice thing happens. Suppose you are planning to excavate two tunnels for the arms of your 60 degree interferometer. By increasing the excavation budget by only 50 percent, you can dig a third tunnel to complete the triangle. Then, by installing an interferometer at each vertex of the triangle, as shown in Figure 9.3, you can *triple* the number of interferometers you can install. The loss in sensitivity in going from 90 to 60 degrees is easily compensated by having three interferometers running simultaneously in the same place. In the current design for ET, each side of the triangle is to be 10 kilometers long, as compared to 3 or 4 kilometers for the existing interferometers. The increase in arm lengths exploits the fact that the change in separation between two objects that a gravitational wave induces is proportional to the distance between them. Moreover, the tunnels will be 100 to 200 meters below the Earth's surface to reduce seismic and gravity noise, and the mirrors will be cooled to temperatures near absolute zero to reduce thermal noise. Three detectors allow us to measure gravitational waves with increased sensitivity arriving from all directions, and they make it possible to separate the different polarizations of the waves at a single site. To be brutally realistic, today's vision of ET is so expensive that it is likely that only a single version will be built. So ET is designed to extract the most science possible if one is limited to a single high-sensitivity instrument.

Third-generation gravitational wave detectors such as ET will inaugurate the era of *precision* experimental relativity in extreme gravity.

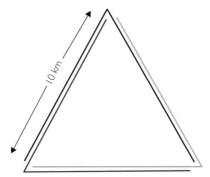

Fig 9.3 Three-interferometer concept for the Einstein telescope. In an equilateral triangular tunnel, three separate interferometers can be installed. The proposal calls for arms 10 kilometers long.

The Einstein Telescope's increased sensitivity will allow it to hear signals from even farther away than any other ground-based detector. For example, ET is expected to detect essentially all mergers of stellar-mass and intermediate-mass black holes, as well as neutron stars, anywhere in the observable universe. Moreover, the events that advanced LIGO and Virgo have already detected would have been a hundred times louder, had they been detected by ET. Such increased sensitivity will allow us to dig deep into even the smallest features of the waves, enabling us to extract tiny details of the source that generated them. In addition, we will be able to carry out tests of general relativity that are at least a factor of a hundred more stringent than current tests.

Still, despite the most visionary design concepts, all of these detectors have the same limitation. At low frequencies, an interferometer's output is dominated by seismic and gravity noise, even when it is buried deep underground. The second and third generations of detectors may allow us to detect waves with frequencies as low as a few hertz, but to go to even lower frequencies is impossible for a detector on Earth. We need to go to space.

The idea of going to space to detect gravitational waves is not new. In 1976, a NASA sub-panel on relativity and gravitation chaired by Rai Weiss published a report that discussed both ground-based laser inteferometers and the possibility of putting such a system in space.

While that report envisaged a Michelson-type interferometer in orbit, some researchers, such as Peter Bender and James Faller of the University of Colorado, Ron Drever in Glasgow, and others, began to ruminate informally about an array of free-flying spacecraft. In papers published in 1984 and 1985, Faller, Bender and other Colorado colleagues outlined a possible space antenna for gravitational waves.

They imagined putting three satellites into orbit around the Sun, separated by about a million kilometers. The three would be in an L-shape arrangement, just like the ground-based interferometers that were being studied at the time. Laser beams would be used to monitor the separation between the main satellite at the corner and the satellite at the end of each "arm." There would be no beam tubes, of course, since the vacuum of space is far better than any human-made vacuum. The satellites would orbit freely, unencumbered by support wires and unaffected by Earth-bound seismic noise.

However, you cannot send a laser beam a million kilometers, bounce it off a mirror and look for the reflected signal. The reason is that, no matter how well you focus the laser beam, it spreads out, and by the time it reaches the end satellite, the beam is so large that the mirror would reflect only the tiniest fraction. That reflected beam also spreads out, and so by the time it returns to the main satellite, there is effectively no light intensity left to detect. At the 4 kilometer distances of the ground-based interferometers this is not a major problem, but at a million kilometers, forget it. But there was a solution. Instead of reflecting the beam at the end satellites, equip each satellite with its own laser. Then capture the arriving laser beam using sophisticated optical devices, measure the value of the "time stamp" (also called "phase") that was encoded in the beam when it was sent out, and imprint that value in a new and powerful signal emitted by the onboard laser back toward the main satellite. In fact, this kind of scheme is a laser version of the "radar transponder" used to track interplanetary spacecraft very precisely (see Chapter 3).

But even though the vacuum of outer space is good enough for the laser beams to travel freely, it is not good enough for the satellites themselves. Satellites are being continuously buffeted by the solar wind and by the photons of light emitted by the Sun. To some degree, these

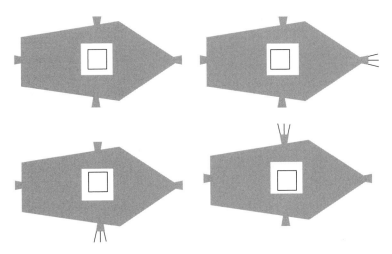

Fig 9.4 Drag-free control in LISA. If external forces push the spacecraft so that the central cube gets too close to a wall of the chamber, signals are sent to micronewton thrusters to nudge the spacecraft to keep the cube centered.

effects are fairly constant in time, and can be modeled and accounted for. But they also fluctuate significantly, leading to a random jiggling of the spacecraft that is every bit as worrisome as seismic noise on Earth. To get around this, Faller and colleagues proposed to use the idea of "drag-free" control, a concept that had been tested using US Navy navigation satellites called TRIAD during the 1970s, and at the time was under active development for Gravity Probe-B (see page 96). The idea would be to have a small mass inside each spacecraft (a cube with reflective surfaces, for example) serve as the "end" of each arm (top left panel of Figure 9.4). The laser beam enters the spacecraft, bounces off a surface of the cube, and then enters a sequence of lenses and mirrors that measure the required information and send that to the onboard laser to generate the return beam.

The crucial thing is to protect that small cube from the buffeting action of the solar wind and the stream of solar photons. The surrounding spacecraft does the required shielding, of course, but those external forces could push the spacecraft so that the cube is no longer centered inside its protective chamber but gets too close to one wall (top right panel of Figure 9.4). Sensors inside the chamber detect this proximity and send signals to thrusters on the opposite side of the spacecraft. These thrusters nudge the spacecraft so that the cube is re-centered. The spacecraft has thrusters pointing in all six directions so that external forces from anywhere can be compensated. In this way, each cube moves on a path provided purely by curved spacetime, and by any ripples in spacetime that pass by in the form of gravitational waves. The typical force needed to do the job is measured in units of a "micronewton." One micronewton is roughly the downward force exerted on your dog's tail when a flea lands on it.

These ideas would ultimately become LISA, the Laser Interferometer Space Antenna, a mission being developed by the European Space Agency, with participation by NASA, and planned for a launch in 2034. We have seen several examples in this book of the often long and tortuous path between inception and fruition of major undertakings in gravitational physics: 43 years between the proposal by Fairbank, Schiff and Cannon and the launch of Gravity Probe-B (Chapter 4); 30 years between Rai Weiss' MIT report and LIGO's first science run in 2002, and 44 years to the first detection. But at 54 years, LISA will set a record. To some, this may seem like an intolerably long time. Some of the originators of the idea may not live to see it finally come to fruition. Some may spend their entire careers working on this one project. But this is part of the landscape of science. When it comes to answering the deepest questions about the universe, scientists often have to play the long game; whether it involves constructing the most energetic particle accelerators, perfecting nuclear fusion, or building space telescopes.

The Faller–Bender concept drifted around the community of researchers for a number of years. An alternative version involving spacecraft orbiting the Earth was put forward by Ronald Hellings of the Jet Propulsion Laboratory. Another person interested in a space detector was Karsten Danzmann of the Max Planck Institute in Garching, Germany, who was engaged in building the GEO-600 ground-based interferometer. In 1992, it all came to a head.

That year, the thirteenth International Conference of General Relativity and Gravitation (the conferences that began with GR0 in Bern in

1955) was held in Huerta Grande, Argentina, a picturesque little town in the middle of the country in the province of Cordoba. The conference venue was one of the many "summer holiday camps" built for the families of workers in labor unions around Argentina's countryside by Juan Domingo Perón during his presidency between 1946 and 1955. At that meeting, there was considerable discussion among Bender, Hellings, Danzmann and other proponents of a space detector, and an agreement was reached to submit a proposal to the European Space Agency, which had put out a call for proposals for new space missions. There is a legend that as the conference came to an end and participants began heading back home to the United States, Europe and Asia, a luggage strike at the airports took place. This caused delays and cancelations of flights, forcing a large group of relativists to be stranded in the middle of Argentina. It was during this forced delay, recall some, that the agreements to make a pitch to ESA were forged. Needless to say, with the passage of time, the memories of the key people have become fuzzy on this point, so we will leave it there as a legend.

By 1997, following numerous assessment studies by both ESA and NASA, a "baseline" mission for LISA was developed (we will describe it shortly). The science of the proposed mission was impeccable, the technical challenges would be great, and the mission would be expensive, with costs likely to be measured in billions of US dollars. But unforseen events and circumstances arose that kept pushing LISA farther into the future. For example, the discovery of extrasolar planets in 1997, and the discovery in 1998 that the expansion of the universe was accelerating rather than decelerating, brought forth numerous proposals for space telescopes to find more planets and to solve this cosmic riddle. And beginning around 2005, NASA's James Webb Space Telescope, being built as the successor to the Hubble Space Telescope, began to experience substantial cost overruns and delays, eating into the funds available for all new projects, such as LISA. By 2010 it was clear that the top priority, billion dollar astrophysics mission that NASA would mount after the Webb telescope would be a mission called the Wide Field Infrared Survey Telescope (WFIRST), designed to probe the acceleration of the universe in more depth and to look for more exoplanets. LISA was third on the list, behind a suite of small missions, known in NASA jargon as "Explorer" class.

Meanwhile, ESA was planning its long-term strategy for large space missions, and convened a meeting in Paris in 2011 to organize the queue. Three large projects were competing to be the first in a proposed sequence of launches, at roughly eight-year intervals, beginning in the early 2020s. They were the Jupiter Icy Moon Explorer (JUICE), the Advanced Telescope for High Energy Astrophysics (ATHENA), and LISA. The presentation made by the LISA team received a standing ovation, but rumors coming out of Washington DC forced ESA to defer ranking the missions.

The year before, the Republican Party had won a majority of seats in the US Congress, and they refused to raise the country's debt ceiling without a negotiation over deficit reduction. On 31 July 2011 the Congress agreed to raise the debt ceiling in exchange for severe spending cuts in the future. These cuts, together with the James Webb telescope overruns, hit NASA hard, forcing it to freeze development of almost all future missions, including joint projects with ESA, such as JUICE, ATHENA and LISA.

With NASA out of the picture, ESA came up with a compromise: they would go it alone, but the design of each mission would have to be descoped to save money. For two of the missions, descoping was somewhat straightforward. In JUICE, the spacecraft could visit just one of Jupiter's moons, say Ganymede, instead of also visiting Europa. For ATHENA, you could simply remove one of the secondary X-ray telescopes from the satellite. But by this time, the LISA design was for a triangle-shaped interferometer, with three satellites separated by five million kilometers each and with two laser links along each arm of the triangle, very similar in concept to the triangular design of ET shown in Figure 9.3. Descoping this design was not so simple. Decreasing the arm lengths from five to one million kilometers saved money in focusing requirements for the telescopes directing the laser beams, and in the fuel needed to situate the three spacecraft properly, but not enough. So the LISA team also proposed eliminating one of LISA's arms, returning to a V-shaped arrangement, very much like the original

LISA proposals of the 1980s, thus dramatically simplifying two of the three spacecraft.

But ESA still had to decide on the ordering of the launches of each descoped mission, so it called for another "shoot-out" in 2013. This time the LISA proposal ranked first! Everything was set for the construction and launch of the mission, except for one condition: the LISA team had to demonstrate that the technology required for LISA was possible in reality. This forced them to design, construct and launch an entirely separate mission, to be called LISA Pathfinder. Such a requirement is not unprecedented. We recall from Chapter 4 that Gravity Probe-B planned a technology demonstration mission, but the Challenger disaster caused it to be scrubbed. In retrospect, such a "GP-B Pathfinder" might have discovered the "patch effect" problems that caused that project such grief. Ironically, the LISA proponents had proposed a Pathfinder-type mission back in 1998, but it had never been accepted by the agencies.

The idea of LISA Pathfinder was to test the most challenging aspects of LISA, the ability of the optical sensors to make extremely precise measurement of the location of the floating "test" masses, the drag-free control system, and the micronewton thrusters. The experiment consisted of a satellite with two test masses separated by 38 centimeters that represented a miniaturized version of one of the LISA arms. There is no need for large separations because light propagates the same no matter how far it travels. The test masses were made of gold and platinum, machined as perfect cubes of 4.6 centimeters (about 2 inches) on a side and weighing about 2 kilograms. When the satellite was launched, and while it was en route to its final orbital configuration, the test masses were secured inside their chambers to prevent them from bouncing around. But once the satellite reached its final orbital configuration, roughly 74 days after launch, the test masses were released and allowed to float freely as the spacecraft orbited around the Sun. The satellite had accurate sensors to constantly measure the position of each test mass relative to the walls of its chamber. Two different micronewton thrusters were tested, one a device that emits a minuscule stream of cold gas, the other a device that uses electric fields to accelerate a stream of a few ions (recall the flea on the dog's tail).

The eighteen-month mission was launched on 3 December 2015, with the idea of coming close enough to the performance requirements of the full LISA mission that the agency could feel confident that the final goals could be met with subsequent development. Pathfinder succeeded far beyond anybody's expectations. The sensors were able to measure the relative position and orientation of the test masses to better than a billionth of a centimeter, and the thrusters were able to keep the test masses isolated to this accuracy. LISA Pathfinder actually beat many of the performance goals.

In fact, Pathfinder's extraordinary performance was clear within a few months of the launch. Meanwhile, on day 70 of the mission, LIGO announced the first detection of gravitational waves, making gravitational wave astronomy a reality. These two facts led to a dramatic turnaround for LISA's fortunes. In June 2016, a NASA committee recommended ways that the agency could rejoin LISA as a junior partner with ESA. In August, an assessment of US priorities for all of astronomy and astrophysics recommended that NASA restore support for LISA, and a month later a NASA official announced to a symposium of LISA scientists that the agency was ready to return. Finally, in June 2017, ESA announced approval of LISA, with the third arm restored, for launch around 2034.

Figure 9.5 shows what the "final" version of LISA will look like. Three independent spacecraft will orbit the Sun once per year on orbits very similar to the Earth's orbit. If the initial orbits are chosen in a clever manner and there are no disturbances to the orbit, then a miraculous thing happens: the three spacecraft orbit the Sun in formation, maintaining the rigid triangular arrangement with equal separations, without the need to fire their thrusters. The secret is to start the orbits on a plane that is tilted by 60 degrees relative to the Earth's orbital plane.

To get an idea of how this would work, let's assume that the Earth's orbit is perfectly circular (it isn't quite, but that is a detail that can be added later). At the beginning of the orbit, one spacecraft is below the Earth's orbital plane, and closer to the Sun (it is at perihelion), while the other two are above the plane, and farther from the Sun. When the three reach the opposite side of the Sun, the orientation is reversed.

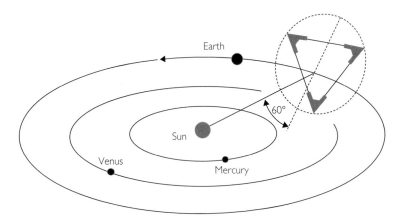

Fig 9.5 Three LISA spacecraft orbit the Sun in a triangular formation. The plane of the triangle is tilted by 60° relative to the plane of the Earth's orbit. Laser beams travel in both directions between the spacecraft, separated by about three million kilometers. The array trails the Earth by about 20°.

The spacecraft that was below and inside is now above and outside (it is now at aphelion), while the two that were above and outside are now below and inside (recall the elliptical orbit displayed in Figure 3.1). But they will have exactly the same separation as they had at the start. Figure 9.6 shows that, as the trio orbits the Sun counterclockwise as seen from above, the triangle rotates clockwise within the 60 degree tilted plane, retaining its equilateral triangular shape at all times. No thrusters are needed to achieved this beautiful formation flying. Newton's theory of gravity (which is a good enough approximation here) does all the work. Probably only a physicist would call such a solution of Newton's equations "cute," but we assure you, it is *extremely* cute.

The approved LISA design calls for arms three million kilometers long. In Figure 9.5 we have blown up the triangle by a factor of about forty-five; in reality it would be the size of a tiny speck on that figure (the Sun and the planets are also blown up way out of proportion). On the other hand, the arms will be about eight times the Earth–Moon distance. The array will follow the Earth in its orbit, lagging behind by about 20 degrees. There is nothing magic about that number. It is determined by requiring that the array be far enough from the Earth–Moon system that the gravitational tug of those bodies doesn't distort the triangle too much, but still close enough that the data can be easily transmitted to Earth. Each spacecraft will have two of everything: two cubic test masses, two lasers and two telescopes. Two laser beams will propagate continuously along each arm, one in each direction. As each beam arrives at each spacecraft, its "timestamp" or phase will be recorded, and these six streams of data will be sent to Earth to be analyzed in search of tiny variations induced by a gravitational wave passing through the array.

Once LISA launches, it will open our ears to a whole new genre of cosmic music. The ground-based interferometers are sensitive to frequencies between about 10 and 1,000 hertz, or wavelengths between 30,000 and 300 kilometers. Because of the seismic noise wall, lower-frequency waves are inaudible to those devices. By contrast, LISA will be able to hear waves between ten millionths of a hertz and a few tenths of a hertz, or wavelengths between 200 astronomical units (the Earth–Sun distance is one astronomical unit) and a few million kilometers. The spacetime ripples detectable by LISA will undulate with periods ranging from 10 seconds to a day. Just as the transition from astronomy using visible light to astronomy using light in the radio band transformed our view of the universe, so too will the low-frequency gravitational wave band reveal a whole new world of exotic sources.

What kinds of sources would they be? The prototypical example is the merger of two supermassive black holes. We learned in Chapter 6 that most massive galaxies, including our own, contain a massive black hole in their center. Their masses range from a million to ten billion solar masses. It is also well known that galaxies can collide with each other and merge. This is not a very common occurrence in the present universe, although it is known that our Milky Way and the nearby Andromeda galaxy are speeding toward each other at over 100 kilometers per second. Whether they will actually collide and merge or pass each other by in more of a "strangers in the night" encounter is a subject of current debate. However, earlier in cosmic history galaxies were much closer together, more likely to tug at each other via their mutual gravitational attraction, in spite of the overall universal expansion, and more likely to merge, and this is borne out by observations.

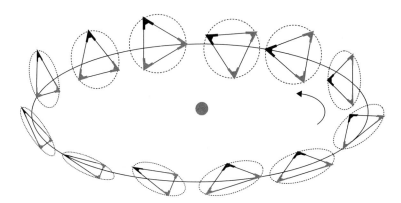

Fig 9.6 As the LISA configuration orbits the Sun, the plane of the triangle rotates, and the three spacecraft perform a cartwheel within the plane. This special orbital solution keeps the distance between the spacecraft approximately constant.

If each galaxy has a supermassive black hole in its core, then the two black holes, being much more massive than stars, will settle toward the core of the merged galaxy and begin to orbit each other. In fact, astronomers have found a few examples of galaxies that show all the signs of being the product of a merger, and that also have *two* very small and very bright spots in their cores, each presumably a supermassive black hole with its accompanying accretion disk of hot gas. When the two black holes finally begin their fatal inspiral dance, they will emit spectacularly loud gravitational wave chirps.

LISA's "ears" will hear the waves far above the instrument's intrinsic noise, quite unlike the chirps detected from black hole inspirals by LIGO–Virgo, which required sophisticated filtering of the background noise (see Figure 8.2). LISA will be able to detect these sources from the farthest reaches of the universe, and because gravitational waves travel at the same speed as light, this means that LISA will detect massive mergers from the very early universe. LISA will be able to answer questions such as: did supermassive black holes form at the same time as the earliest galaxies, or did they form much later? To astronomers who study the formation and evolution of galaxies this is a burning question, and conventional astronomical telescopes cannot "see" far enough to answer it.

Another audible gravitational wave signal is produced when a small black hole or a neutron star falls into a supermassive one. This typically occurs in the dense core of a galaxy when an unfortunate small black hole or neutron star has a close encounter with another body such as a star, and suffers a gravitational "slingshot" that sends it hurtling toward the massive black hole lurking at the center. It would take an incredibly lucky (or unlucky) aim for the body to go straight across the event horizon of the black hole, simply because the black hole is an incredibly small target. The more likely outcome is that the body will come close to the black hole, whirl around it at maybe half the speed of light, zoom outward to a large distance, then fall back toward the hole and whirl around it again and so on. But instead of moving on a nice elliptical orbit, as depicted in Figure 3.1, the body will undergo a very complicated gyrating motion, induced by the strong general relativistic warpage of spacetime near the black hole. Figure 9.7 gives an example of what such a "zoom–whirl" orbit might look like. All the while the orbit will slowly shrink, because of the orbital energy being lost to the gravitational waves being emitted. Because the ratio of the mass of the small black hole or neutron star to that of the supermassive black hole is extremely small, this is called an EMRI, or extreme mass-ratio inspiral. EMRIs produce gravitational waves that are very rich in frequency structure. Close to the black hole, the body's motion is changing very rapidly, so the waves emitted have high frequencies, while far from the hole, the motion is changing slowly, so the waves have low frequency. If we played these gravitational wave sounds on a speaker, they would sound like a throttling motorcycle, with the bumps and peaks associated with the many frequencies present in the EMRI orbit.

A third source of gravitational waves that will become audible with LISA are called "galactic binaries." These are binary systems of two white dwarf stars, objects that sometimes result from a supernova and are supported from gravitational collapse by a special type of quantum mechanical pressure (see Chapter 2). Recall that these objects have masses that are typically between a tenth and one and a half times the mass of the Sun, but they are much smaller than the Sun in size, with radii of several thousand kilometers. White dwarves in a binary system

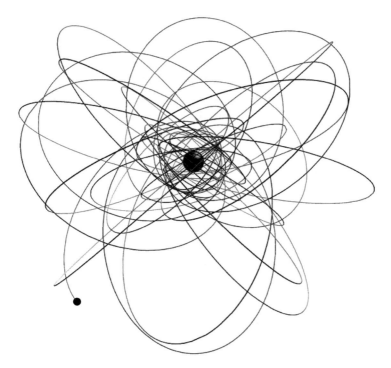

Fig 9.7 Zoom–whirl orbit of an extreme mass-ratio inspiral around a spinning black hole. Because of the large pericenter advance and the strong dragging of inertial frames, the orientation of the elliptical orbit undergoes wild gyrations, while at the same time shrinking because of gravitational radiation emission.
Credit: Maarten van der Meent.

are still far away from each other, so that the gravitational wave signal is very weak and the inspiral is negligible. This implies that the frequency of the waves stays constant over long periods of time. But for many of them the frequency of the gravitational waves is in the millihertz band, right where LISA is most sensitive.

Because the waves are weak, we can essentially hear only those white dwarf binaries within our own Milky Way galaxy. About ten such binaries are known from electromagnetic observations, but because white dwarfs are intrinsically dim, these are all in the immediate solar neighborhood. It is not known how many there are altogether in the galaxy but there could be millions (the galaxy contains a trillion stars in total). In addition, there could be neutron star and black hole binaries in wide orbits, emitting no electromagnetic radiation, but sending out detectable gravitational waves. Learning about these populations could teach us a lot about how stars evolve, age and end their lives as one of these compact objects.

The closest ten or so white dwarf binaries have had their orbits determined very precisely using electromagnetic observations of the stars' motions. As a result, we can predict the gravitational wave signal they emit almost exactly using general relativity. These are called "verification sources," because LISA had better detect them clearly and unambiguously, otherwise there is something wrong with LISA or with general relativity. LISA may also be able to identify up to 25,000 additional binaries, especially at higher frequencies. But there may be so many binaries in the galaxy that they will produce a background of "noise" that is actually louder than the intrinsic noise within the instruments.

This is familiar to anybody who has been in a sports stadium. If you close your eyes and listen, you will hear many voices talking, laughing, yelling at the same time. It is hard to pick up any one voice from the combination. You can perhaps pick out the closest and loudest ones, but that's only a few in a sea of noise. With galactic binaries, we will still be able to pick out a few of the waves that stand out in this background, but in general, the wealth of sources will contribute to a form of noise.

New sources of gravitational waves mean new tests of general gelativity. Probably the strongest advantage of these new sources is that they are so loud that they can be heard even when their distance to Earth is truly cosmologically large. Certain modifications of Einstein's theory, such as the idea that gravitational waves might travel at a speed different from that of light, or that the particle associated with gravity (the graviton) might not be massless, have effects that build up with the distance traveled. Therefore, LISA observations of supermassive black holes will lead to tests of such deviations that could be a million times more powerful than what has been achieved by the ground-based interferometers (see pages 228 and 239).

In an extreme mass-ratio inspiral, the small black hole acts as a tracer of the warped spacetime geometry around the spinning hole: now very close, now very far away, now on the equatorial plane, now over one of the poles, mapping space and time the way a cartographer documents every bump and every valley on the Earth's surface. These inspirals can trace the gravitational field outside of a black hole similar to the way GRACE traces the gravitational field of Earth (see Chapter 4). The gravitational waves emitted encode all of this information, and the detection and analysis of the waves then allows us to determine whether the geometry of a supermassive black hole is truly described by Einstein's equations or not.

And there is one more surprise that LISA has up its sleeve: it allows for multi-wavelength observations. Imagine that LISA observes the gravitational waves emitted by a binary composed of black holes, each with a mass of roughly 30 solar masses. For LISA to achieve this, the binary has to be relatively close to Earth, say 1.4 billion light years away. In addition, the orbital separation of the binary has to be large enough that it is emitting gravitational waves at a sufficiently low frequency that LISA can hear them. Such a source would be in LISA's sensitivity band for many months, with its frequency slowly increasing and the black holes slowly spiraling into each other. Eventually, the waves' frequency will increase beyond the detector's sensitivity band. This is similar to listening to a sound whose frequency is slowly getting higher and higher, until eventually you can't hear it anymore because our ears cannot detect sound above a certain frequency.

Such a signal would then disappear from LISA's band, but a few months later the same signal, now at a frequency of 10 hertz, would reappear in the ground-based detectors! We call such an observation, first by LISA and then by ground-based instruments, a multi-wavelength event because the same signal is observed at very different frequencies by different instruments.

The potential of such a multi-wavelength event is truly breathtaking. The LISA observation would make it possible to predict when in the future the same event would be detectable by ground-based instruments. Moreover, LISA would be able to localize the source in the sky, thus allowing telescopes on Earth and in orbit to point to this box in the sky to watch the fireworks (if there are any) as they occur. And for tests of general relativity, such an event would also be fantastic, since it would allow us to measure very precisely the rate at which the frequency of a binary changes over many, many orders of magnitude. This rate of change is very precisely predicted in general relativity, so any deviation from these predictions would be catastrophic for Einstein's theory.

There is a final advantage of LISA for testing Einstein's theory. Many of the sources that LISA will see will be observable for many months, and even years, in contrast to the short chirps of minutes to fractions of a second that LIGO–Virgo detected. During this time the LISA triangle will be in orbit around the Sun, with both the orientation and the plane of the triangle rotating around during the orbit. In this way, LISA will behave like the multiple interferometers on Earth, allowing us to extract the different polarizations of the signal, even though we only have a single instrument. Therefore, LISA can test Einstein's prediction that gravitational waves only have two polarizations, the "plus" and the "cross" modes shown in Figure 7.3. This is a strong statement because many modified theories predict additional polarizations. If we can verify that gravitational waves truly only possess two polarizations, it would be a fatal blow to many modified gravity theories.

So far we have discussed in detail the many different instruments we can build to detect gravitational waves, but there is a natural detector that is provided to us by nature; it involves pulsars (see Chapter 5 for an account of the discovery and characteristics of pulsars). The pulses emitted by some of these objects, particularly the old, recycled pulsars, are incredibly stable over long times, arriving at the detector with essentially the same pulse period week after week, year after year. Most of these special pulsars have rotational periods of milliseconds, so the neutron stars that produce the radio beacons are spinning faster than the blades in a professional kitchen blender. Their periods are typically stable to better than a microsecond over a year, making them competitive with some of the best atomic clocks on Earth. But if a gravitational wave goes through spacetime between the pulsar and Earth, their measured periods will be altered by the passing gravitational waves. One can think about it crudely as the wave alternately stretching and shrinking the distance between the Earth and the pulsar, thus inducing an increasing and decreasing Doppler shift in the apparent frequency of the pulses.

Therefore, a sufficiently precise measurement of the pulse frequencies could in principle be used to detect gravitational waves.

One problem is that although millisecond pulsars are stable over long times, they are not so stable over short times, say of the order of tens of milliseconds. This is because of the intrinsic variability in the pulse frequency, but also because the speed of the light pulses varies as the radio signals propagate through the clumpy ionized gas, called the interstellar medium, that lies between us and the pulsar. Avoiding such speed variations in the light signals is why LIGO–Virgo had to have ultra-high vacuum tubes. So the observed pulses need to be averaged over long periods, of the order of a year, in order to averge out any intrinsic or propagation-induced variability. This is similar to the fact that, over ten billion flips of a perfect coin, the outcome will be fifty–fifty heads vs. tails to one part in 100,000, while over ten flips you could easily get four heads in a row, followed by three tails and then two more heads. The result of such averaging is a set of pulses with the remarkably stable period described above. On the other hand, when a given gravitational wave from a very distant source passes through the Milky Way, it affects the measured variations in pulse period from a given pulsar in a very predictable way that depends only on the direction to the pulsar relative to the direction of the source of the waves, and on the distance between the Earth and the pulsar. Therefore, if one can observe pulses simultaneously from an *array* of pulsars, say fifteen or twenty, distributed all around the Milky Way, then it is possible to do a better job of averaging out the short-term fluctuations in pulse periods that are specific to each pulsar and are more or less random, revealing the common gravitational wave signal.

This effort, known as pulsar timing arrays (PTAs), is being carried out by teams of radio astronomers around the world. There is the Parkes Pulsar Timing Array (PPTA) based at the Parkes radio telescope in Australia, the European Pulsar Timing Array (EPTA) that uses the four largest radio telescopes in Europe, and the North American Nanohertz Observatory for Gravitational Waves (NANOGRAV) that uses the Arecibo and Green Bank, West Virginia telescopes. An international organization (IPTA, what else?) oversees and coordinates the various teams.

What kind of gravitational waves can we detect with such PTAs? Since we have to average the pulses over a year, then we can only detect waves that vary over longer times. Since a year is about 0.03 billion seconds, we are thus looking for waves with frequencies of around 30 billionths of a hertz or 30 nanohertz. At these frequencies, the loudest sources of gravitational waves are orbiting supermassive black holes in the centers of galaxies, but long before they are close enough to each other to merge. There could be thousands to millions of potential sources in the sky at nanohertz frequencies. So many sources cannot necessarily be separated in the data, but instead they lead to a noise background, similar to the white dwarf binary background for LISA. But this "noise" has very specific characteristics that can be used to distinguish it from other noise sources, such as intrinsic pulsar jitter and propagation effects. By focusing on this characteristic (called a correlation), we can learn about the black hole binary population that produced the waves.

A background of gravitational waves from binary massive black holes has not yet been detected by the PTA method, but detections may be imminent. Once this is achieved, we will be able to learn not just about the gravitational wave background itself, but also about other astrophysical effects that may be causing the supermassive black holes that produced the waves to inspiral toward each other. And when this is taken into account, one can also in principle place constraints on modified theories of gravity that predict the existence of other gravitational wave polarizations, because the "correlation" we mentioned depends on the number of polarizations in the waves.

As we have seen, gravitational waves have provided a soundtrack to the universe, and the sophisticated headphones we have constructed to hear it will only improve in the coming decades with third-generation ground-based detectors, LISA, and PTAs. Einstein's gravitational wave playlist encompasses many keys and genres, from the high-pitched, staccato chirps of the stellar mass inspirals to the hiss of the white dwarf binary noise to the motorcycle vroom of EMRIs. All we have to do is listen.

CHAPTER 10

A Dialogue

Nico Cliff, you've been working in general relativity since like a million years before I was even born. You must have had some pretty exciting moments during your career. What was number one?

Cliff It has to be learning that gravitational waves had been detected by LIGO. We were all pretty confident that the advanced interferometers would eventually detect waves, but we had no idea when. As you know, during the fall of 2015 there were many rumors going around that LIGO had something, but since I was not part of the LIGO–Virgo collaboration, I had no inside knowledge. Finally, in late January 2016, about two weeks before the actual announcement, various science journalists started calling me to ask if I knew anything. All I could say was that I had heard the same rumors as everybody else. Finally, I emailed some colleagues in my department at the University of Florida, who *were* members of the collaboration, if I should begin thinking about preparing some remarks for journalists. One of them answered "you should always be prepared for whatever life sends your way." This was not very helpful! But a few days later, another colleague came into my office, closed the door and made me swear not to reveal what he was going to show me, which was the draft of the discovery paper. Needless to say I was blown away! I'm normally a very calm person, but when I saw my wife that evening, she immediately said, "What's wrong?" I had to tell her! (I knew she could keep the secret.)

A week later came the big press conference. In many ways, my career began in 1969 with Weber's flawed claims of detection of gravitational waves, so I'd been waiting almost fifty years for that moment.

I have a feeling that your most exciting moment might be the same, but how did you learn about it and what was your reaction?

Nico I was in the same boat as you. Back when I was a graduate student at Penn State, I was a member of the LIGO collaboration. But then, after graduate school, I decided I was more interested in theoretical physics, so I stepped away, although my research has always remained very close to the science that one could do with LIGO observations. By the fall of 2015 I was already at Montana State, and although colleagues of mine were members of the collaboration, I wasn't, so I was not privy to any "secret" information. But I was hearing the same rumors as everybody else. By the time of the announcement, everybody in our institute essentially knew what was happening, so we gathered in the conference room and bought a cake to celebrate when those five magical words were uttered: "We have detected gravitational waves."

I was a bit surprised when I heard that one of my colleagues told his ten-year-old son about the discovery essentially the day it was made, back in September 2015. But the boy kept the secret! Now, we get alerts on our mobile phones. I'm happy we are finally done with the secrecy. I'll bet this is nothing like what the field was like back when you were young, you know, back in the hippie era.

Cliff Indeed, my hair was appropriately long, particularly compared to today. That reminds me of another exciting moment from "back in the day." In the fall of 1974 I was a brand new assistant professor at Stanford University, working with the astrophysicist Robert V. Wagoner. In late September, Bob burst into my office waving an "IAU telegram."

Nico A telegram! Really?

Cliff In those pre-internet days, new discoveries in astronomy were announced by a service of the International Astronomical Union, which would send a telegram to observatories and universities world-wide outlining the basic facts of the discovery. Bob, who is one of the most enthusiastic physicists I know, shouted "Cliff, they've just discovered a pulsar in a binary system! Whatever you are working on, drop it. We have to get to work on this new system." Of course, this was the Hulse–Taylor binary pulsar, which we discussed in Chapter 5. Within a few weeks, Bob wrote a key paper on how it would be possible to measure the change in the orbital period of the binary and thus prove the existence of gravitational waves, and I wrote a paper on the implications of measuring the advance of the pericenter of the orbit. For the next eight or so years, the science related to the binary pulsar occupied almost half of my research life. And when Joe Taylor announced at the 1978 Texas Symposium his team's measurement of the inspiral of the orbit in agreement with Einstein's prediction for gravitational wave energy loss (page 128), it was a fantastic moment. Of course, the icing on the cake came in October 1993, when I received a fat envelope from the Nobel Foundation containing an invitation for me and my wife to attend the Nobel Prize ceremonies in Stockholm honoring Joe and Russell Hulse.

Nico Wow! Obviously, there have been many moments during your career. Another big moment for me was when Frans Pretorius completed the first computer simulation of the merger of two black holes in full general relativity. When my physics life started at Washington University in Saint Louis, in your gravity group, if you recall, nobody could simulate such a collision. It was really hard even to use a computer to simulate a single black hole just sitting there doing nothing (even though we had the Schwarzschild solution, which you can write down in one line on a piece of paper). One of the problems was the apparent singularity at the event horizon of the black hole. Even though some mathematical functions become infinite there, we know how to handle that and we know that nothing physically bad happens there.

Cliff Indeed, we discussed this in Chapter 6. But while our minds can fathom infinity and even make peace with it, computers can't; they simply stop and emit an error message whenever they encounter a number divided by zero. And this was only one problem with computer solutions of Einstein's equations.

Nico But it all changed in 2005. I was a graduate student at Penn State at the time, and I recall being in my office, because I was too young to attend a conference in Banff, Canada, that more senior students were attending to present their work. Suddenly, a graduate student or postdoc, I can't remember, barged into my office and excitedly told me about the computer simulation that Frans Pretorius had presented in Banff. My jaw dropped! He had done it! It would now be possible to predict what the gravitational waves produced in the final part of a black hole merger looked like. It was the biggest breakthrough I had ever experienced in relativity, and I was super excited. Soon, many other groups around the world were able to merge black holes on the computer, using techniques quite different from Frans', but getting the same result. The black hole merger problem was basically solved.

Cliff That breakthrough came at just the right time, because LIGO was laying the groundwork for advanced LIGO, and this gave confidence that we would have good theoretical predictions for the waves by the time real waves were detected.

Nico Without that 2005 breakthrough we would not have been able to interpret the signals shown in Figure 8.2 so clearly and confidently as coming from a binary black hole merger, and we would not have been able to measure the masses and the distance so accurately.

Cliff By the way, 2005 was quite a year. It was also the "World Year of Physics," when physicists worldwide celebrated the centenary of Einstein's "miracle year" of 1905, when he wrote five papers that transformed physics. Two were on special relativity, one on the photoelectric effect (for which he won the Nobel Prize in 1921), one on the quantum nature of light, and one on Brownian motion. During that year I even went on a four-week, twenty-one-city speaking tour of Canada sponsored by the Canadian Association of Physicists, giving a public lecture entitled (what else?) "Was Einstein Right?" For part of the tour I was accompanied by the Borealis String Quartet, a renowned ensemble based in Vancouver, who performed as a warm-up for my lecture. They played some Mozart (Einstein's favorite composer) and a piece called "From Water to Ice" composed specially for the tour by a University of Alberta physics and music student.

Nico Awesome! I always think it is cool to combine the physics of Einstein with the arts, and we did some of that in 2015 at Montana State for the celebration of the centenary of general relativity. We created an immersive art installation about a black hole with an accretion disk, with some cool spacey music, and also an original interactive movie about Einstein and his theory, in which at the end of the movie the public could generate their own "gravitational waves" by waving a hollow plastic tube (about two feet long) over their heads. Fortunately, people didn't whack each other on the head with the "gravitational wave generators" we gave them! Since then, we've also created a pretty cool spoken word event that combines poetry and physics, as well as a planetarium show about gravitational waves.

Getting back to things we've experienced (and I'm not rubbing it in, or anything), when you were starting out, general relativity as a field was just emerging from the doldrums. How have things changed?

Cliff An obvious change is the growth of the field. Back then there were only a dozen or so relativity "groups" around the world. Every few months we used to get (by snail mail, of course) a list of newly published papers in the field sent from the headquarters of the general relativity society in Bern, Switzerland. The list contained maybe thirty titles. Today, the online archive for general relativity where we post our papers has thirty papers *per day*!

Nico And that number continues to grow. Also, many papers having to do with general relativity are sent directly to the high-energy theory archive or the astrophysics archive, because the authors feel that the papers are of interest to those communities. The overlaps and interactions between general relativity and other fields of physics are really amazing. We mentioned the first Texas Symposium in 1963 (Chapter 6) as an awkward first step to bring relativists and astrophysicists together, but man, are they together now! The neutron star merger paper had 3,000 relativity and astrophysics authors, ten times the number who attended that Texas Symposium. And topics like quantum gravity, string theory and dark energy attract researchers equally from general relativity and particle physics.

Cliff Another big change in general relativity is the role of experiment.

Nico I recall from page 197 the phrase that was used to describe the early days of the field: a "theorist's paradise and an experimentalist's purgatory." Without experimental data, scientists don't take you seriously. Is this how you felt?

Cliff I was lucky enough to begin my career right when this was all changing. What made it exciting was that you could play with lofty spacetime concepts and at the same time "get your hands dirty" with actual data. For instance, my notes from my Caltech graduate student days include mathematical calculations on the motions of bodies predicted by general relativity as well as by other theories of gravity. But from the same period, I also have notes from discussions I had with scientists from the nearby Jet Propulsion Laboratory on what radar tracking accuracy NASA would provide on the proposed Viking mission to Mars, so that they could test the Shapiro time delay effect. As we saw in Chapter 3, that mission was a great success. As time went on, new experimental tests, the development of gravitational wave detectors, and observations in astronomy and cosmology made experiment an increasingly important aspect of general relativity. Today, there is a real symbiosis between theory and experiment, and that makes the field exciting and vibrant, a far cry from the "low water mark" days.

Nico And I guess that's how physics advances. We speculate, and we then build a theory to cement this speculation. But it is not until experiments are done and data is analyzed that we can really trust our theory. And if the predictions of the theory don't match the data, we then discard it, until we find a theory that works. But it's got to work for all the experiments ever done, for all of the data, and then it must survive future tests. It's just amazing to me how a single theory, Einstein's, has managed to do this and remain right for over a century. And it's not as if physicists haven't tried to come up with other theories that could supersede Einstein's and maybe explain some of the anomalies we discussed in Chapter 1. Yet these theories, at least the ones that are able to make predictions, tend to disagree with one observation or another. Experiment seems to really like Einstein's theory, in spite of how crazy and wacky it seems. This is maybe what I find most amazing about general relativity.

Cliff I agree. Any physicist of the 1920s or 1930s would recognize that the general relativity theory we have discussed in this book is *exactly* the same as the general relativity that Einstein presented in November 1915. But those same physicists would be astonished and maybe mystified by today's picture of quarks, leptons, gluons, photons, neutrinos and Higgs bosons that are the fundamental building blocks of matter, and the fundamental theories, quantum electrodynamics, quantum chromodynamics and electroweak theory, that govern them. It is so radically different from the model of their day in which the world was made of electrons, protons and neutrons, all governed by electromagnetism and simple quantum mechanics, that they might think they had been dropped into an alternate universe. These radical changes in the theory of the fundamental particles were driven mostly by experimental results that revealed anomalies or new phenomena that had to be addressed by new or updated theories.

Nico And if we finally detect the particle that makes up dark matter, even more dramatic changes in those theories may be necessary.

Cliff Absolutely. Yet from another point of view, I think maybe the longevity of general relativity is not so surprising. Newton's theory of gravity held sway for over 230 years until general relativity came along. So by that standard, Einstein's theory is still a youngster, like you, Nico. The problem is that gravity is so darn weak. The gravitational force between two protons is more than a trillion trillion trillion times smaller than the electric force between them. So you have to work much, much harder to find the tiny deviations in gravity that might announce a modification of general relativity. So, while the timescale for seeing changes in elementary particle theory may be measured in decades—during my own time as a physicist I've been witness to many big changes in that field—perhaps the timescale for seeing changes in our theory of gravity is measured in centuries.

Nico Well, that's disappointing! I feel like you just dumped a bucket of cold reality on my head! But I suppose you are right.

Cliff There's a legend in our field that I've never been wrong.

Nico True. I even lost a bet with you and had to fork over an expensive bottle of Malbec wine. But we digress. It may well take a very long time to see a big change in our understanding of gravity, to see another scientific revolution *à la* Einstein. It's disappointing because what got me interested in relativity in the first place was *Star Trek*! Not so much all the action with the phasers and photon torpedos, but rather the possibility that new technology would allow us to find loopholes in the current laws of physics and let us explore new worlds and new civilizations. Our current understanding of Einstein's theory suggests that there are no such loopholes, and without them it'll be impossible to witness a black hole from up close, as in *Interstellar*, or to visit those cool new worlds!

Cliff But isn't the world we've got, with its whirling neutron stars, crashing black holes and runaway cosmic expansion, exciting enough for you?

Nico Of course it is! Especially now that we have gravitational wave data! But it's one thing to see the action on the battlefield from up on the hills, like an imperious general on an old horse, and another thing to be up close on the battlefield itself, swinging your sword left and right. It's like the difference between reading about an amazing basketball game and being there with court-side seating! Sure, the event is the same, but the information and the experience are totally different.

Cliff Fair point . . . but I still prefer to be on Earth rather than close to a black hole. On the other hand, if I spent enough time there, when I returned to Earth I could be younger than you! By the way, I'm now curious: are you talking about the original *Star Trek* with Captain Kirk? That show is way before your time!

Nico Yes, I have seen the original series, but I was referring to *Star Trek, The Next Generation*. Back in Argentina, where I grew up, they would show reruns of this show every week on cable television. So by the time I moved to the United States to study physics in 1999, I already knew that black holes were the thing I wanted to study the most. It took me about a year to realize that to study black holes what I really needed to study was Einstein's theory!

Cliff Yes, you obviously got interested early, because I remember that already as a sophomore at Washington University in 2000 you wanted to do research in the topic, first with our colleague Matt Visser, and then with me.

Nico What about you? What got you interested in general relativity research?

Cliff My path was about as different from yours as it could be. At my undergraduate school, McMaster University in Canada, there were no courses in general relativity, and no professors who knew anything about it (this was around 1967). I had read a few popular books and *Scientific American* articles about Einstein and his theories, but that was it. I arrived at Caltech in the fall of 1968 knowing only that I wanted to do theoretical physics, but I had no idea what kind. Many new Caltech grad students in those days wanted to work on particle physics with Richard Feynman, but didn't realize that by that time he was not taking any new students. Some fellow Canadian students told me that I should talk to this brand new professor with a funny name, "Kip." I had never heard of him because he was so new that the brochure that Caltech had sent me describing their graduate physics program didn't list him! Finally, a month and a half into the semester I approached him. He told me I was probably out of luck, because he was teaching the general relativity course as we spoke, but it would not be taught the following year. So I quickly dropped an astronomy elective course and signed up for Kip's course, but I had to catch up on almost half a term's worth of material and homework sets. It was exhausting! But it all worked out, because by the end of that first year Kip invited me to start attending his research group meetings, and the rest is history!

Nico But what about the future of general relativity? Before I moved from Montana State to the University of Illinois I was giving a public talk at the Museum of the Rockies in Bozeman, and an audience member asked me if I "believed" in string theory. I was a bit shocked at first because, as you know, we don't use the word "believe" in physics. I don't "believe" in gravity; I just grab a rock and let it go, and it falls down. Gravity just is, whether I "believe" in it or not. As my students say: "Gravity doesn't care what you think." So after explaining this as politely as I could, I still had to address the elephant in the room:

what about string theory, or other theories of quantum gravity for that matter?

I'm a bit torn about this because we all understand the limitations of Einstein's theory. As we discussed in Chapter 1, we don't really understand what dark matter is or what dark energy is, and they make up 95 percent of the energy and matter in the universe. We suspect that the singularities in Einstein's theory, like the one at the very center of a black hole, are a sign of something missing in the theory. And this is intimately tied up with the fact that quantum mechanics and Einstein's theory are just incompatible at a mathematical level. Some other more "fundamental" theory must replace Einstein's at the very small scales. This more fundamental theory has got to be compatible with quantum mechanics.

So, sure, something like string theory or maybe loop quantum gravity has got to be right. But which one is the right fundamental theory? What I "believe in" or like or find aesthetically pleasing should not matter one bit. What should matter, what should help me decide, is experiment and observation. If there is a quantum gravity theory that makes predictions and these predictions match what we observe in nature, then that's it. We are done! But that's just not where we are at present. None of the quantum gravity theories that are out there are sufficiently complete and devoid of mathematical problems that we can make unique predictions to compare against experiments. But even if they were, the effects of quantum gravity in the regimes where we can make observations today are so minuscule that they seem impossible to measure with today's (or tomorrow's) technology. Am I being too pessimistic?

Cliff Nico, you strike me as someone without a molecule of pessimism in your bones! With so much exciting science to be done with gravitational waves, black holes and neutron stars, there's tons to be optimistic about.

We titled this book *Is Einstein Still Right?*, and we will leave it to our readers to make up their own minds. But another question is: "Will Einstein always be right?" Given my seniority and the fact that I have nothing to lose, I'm going to go out on a long limb here.

Nico Are you saying you are prepared to wager a bottle of wine?

Cliff Sure, why not? Here goes: It would not surprise me if the solution to the universal acceleration turned out to be simply Einstein's original cosmological constant, a constant of nature like Planck's constant or Newton's constant of gravitation, which we have now been able to measure. And it would not surprise me if general relativity turned out to be absolutely correct according to any future experiment accessible to humankind.

Of course, having made that statement, I am reminded of the story about Yogi Berra, the famous New York Yankees baseball player and manager (one of my childhood heroes), and an individual blessed with the most sublimely illogical mind. At some point after his retirement from baseball, his wife Carmen asked him: "Yogi, you were born in St. Louis, you played baseball in New York and we now live in New Jersey. If you should die first, where would you like me to bury you?" Yogi's answer: "Surprise me!"

SUGGESTIONS FOR FURTHER READING

Alexander, Stephon. *The Jazz of Physics: The Secret Link Between Music and the Structure of the Universe*, Basic Books (2017).

Bartusiak, Marcia. *Einstein's Unfinished Symphony: The Story of a Gamble, Two Black Holes, and a New Age of Astronomy*, Yale University Press (2017). New edition of a classic 2000 book on gravitational waves, updated to include the LIGO detections.

Bartusiak, Marcia. *Black Hole: How an Idea Abandoned by Newtonians, Hated by Einstein, and Gambled on by Hawking Became Loved*, Yale University Press (2015).

Binétruy, Pierre. *Gravity! The Quest for Gravitational Waves*, Oxford University Press (2018). Gravitational waves, including the recent detections, by a leader of the Virgo project.

Crelinsten, Jeffrey. *Einstein's Jury: The Race to Test Relativity*, Princeton University Press (2006). Historical account of the early attempts to test general relativity.

Kennefick, Daniel. *No Shadow of a Doubt: The 1919 Eclipse That Confirmed Einstein's Theory of Relativity*, Princeton University Press (2019). A look back at the famous eclipse measurement.

Kennefick, Daniel. *Traveling at the Speed of Thought: Einstein and the Quest for Gravitational Waves*, Princeton University Press (2007). A history of physicists' efforts to determine whether gravitational waves are real, including the inside story of the Einstein–Rosen "disproof."

Levin, Janna. *Black Hole Blues and Other Songs from Outer Space*, Anchor Books (2017). Black holes and gravitational waves, including the first detections.

Thorne, Kip S. *The Science of* Interstellar, W. W. Norton (2014). An account of Kip's adventures in Hollywood, and of the science (and the scientific compromises) that went into the movie.

Thorne, Kip S. *Black Holes and Time Warps: Einstein's Outrageous Legacy*, W. W. Norton (1994). The science and history of black holes and gravitational waves, pre LIGO.

Tyson, Neil deGrasse. *Death by Black Hole: And Other Cosmic Quandaries*, W. W. Norton (2007).

INDEX